THE WORLD'S CLASSICS

THE OXFORD SHAKESPEARE

General Editor · Stanley Wells

The Oxford Shakespeare offers new and authoritative editions of Shakespeare's plays in which the early printings have been scrupulously re-examined and interpreted. An introductory essay provides all relevant background information together with an appraisal of critical views and of the play's effects in performance. The detailed commentaries pay particular attention to language and staging. Reprints of sources, music for songs, genealogical tables, maps, etc. are included where necessary; many of the volumes are illustrated, and all contain an index.

STEPHEN ORGEL is the Jackson Eli Reynolds Professor of Humanities at Stanford University. His books include *The Jonsonian Masque, The Illusion of Power,* and *Inigo Jones: The Theatre of the Stuart Court* (in collaboration with Sir Roy Strong). He has edited Ben Jonson's masques, Christopher Marlowe's poems and translations, the Oxford Authors *John Milton* (in collaboration with Jonathan Goldberg), and Shakespeare's *The Tempest* for the Oxford Shakespeare. His edition of *The Winter's Tale* is forthcoming.

Oxford Shakespeare titles available in paperback:

The rest of the plays and poems are forthcoming.

WILLIAM SHAKESPEARE

THE TEMPEST

Edited by
STEPHEN ORGEL

Oxford   New York
OXFORD UNIVERSITY PRESS

Oxford University Press, Walton Street, Oxford OX2 6DP

Oxford New York
Athens Auckland Bangkok Bombay
Calcutta Cape Town Dar es Salaam Delhi
Florence Hong Kong Istanbul Karachi
Kuala Lumpur Madras Madrid Melbourne
Mexico City Nairobi Paris Singapore
Taipei Tokyo Toronto

and associated companies in
Berlin Ibadan

Oxford is a trade mark of Oxford University Press

First published 1987 by the Clarendon Press
First published as a World's Classics paperback 1994

British Library Cataloguing in Publication Data
Data available

Library of Congress Cataloging in Publication Data
Shakespeare, William, 1564-1616.
The tempest.
(The Oxford Shakespeare)
Includes index.
I. Orgel, Stephen. II. Title. III. Series:
Shakespeare, William, 1564-1616. Works. 1982.
PR2833.A2074 1987 822.3'3—dc20 86-2568
ISBN 0-19-812917-3
ISBN 0-19-281450-8 (pbk.)

7 9 10 8

Printed in Great Britain by
Mackays of Chatham,
Chatham, Kent

PREFACE

It is a pleasure to acknowledge first my deep indebtedness to Frank Kermode's excellent Arden edition, which, despite a number of basic disagreements, I still find an indispensable text. My survey of the play's stage history took shape under the expert guidance of David Kastan, who generously placed his notes at my disposal. John Bender called my attention to the Henry Peacham emblem of the royal mage and to Stephen Batman's note on the identity of Carthage and Tunis. Students from my seminars at Johns Hopkins and the Folger Shakespeare Library have helped immeasurably to bring my sense of the play into focus: I should single out, for help on particular points, Laura Levine, Alexandra Halasz, Mark Rasmussen, Beverly Hart, and Mark Reckson. For references and valuable suggestions, I am indebted to Stephen Greenblatt, Sir Roy Strong, Nancy Wright, and R. A. Foakes. Some of the material in the Introduction has appeared in my essay 'Prospero's Wife', *Representations*, 8 (1984). The patience, intelligence, and helpfulness of the members of the Shakespeare department of Oxford University Press seem to me beyond praise, and I am especially grateful to John Jowett, who gave my text and commentary a detailed and acute reading, and many of whose suggestions I have adopted. Finally, Jonathan Goldberg read the whole manuscript, listened, discussed, argued, and always encouraged. This book is for him.

STEPHEN ORGEL

CONTENTS

Contents

LIST OF ILLUSTRATIONS

INTRODUCTION

Beginnings and Issues

First Appearances. *The Tempest* stands first in the 1623 Folio of Shakespeare's works. Whether it was placed there by its editors or its publishers, and whatever their reasons,[1] the decision has profoundly affected the play's critical history. It has been taken to imply that the play is an epitome of Shakespeare's career, or of human experience; that it was Shakespeare's valediction to the stage and his last play (though never, more logically, that it was his first); that it was the truest expression of Shakespeare's own feelings, and that in the magician-poet Prospero he depicted himself.

Another historical fact, doubtless more fortuitous, has conditioned views of the play in this century. The two earliest surviving records of productions—the only ones in Shakespeare's lifetime—are of performances at court. 'Hallowmas nyght', according to the Revels Accounts for 1611, 'was presented at Whitehall before the kinges Maiestie a play Called the Tempest',[2] and a year and a half later the play appears in a list of fourteen performances at court during the festive season preceding the marriage of James I's daughter Elizabeth to the Elector Palatine.[3] These records, combined with the fact that the play includes a masque, have seemed to many modern critics to link *The Tempest* specifically with the Jacobean court. In this reading, Prospero becomes not Shakespeare but King James, or a union of the two, and the betrothal masque obliquely celebrates the forthcoming royal wedding. The latter claim requires the further assumption that what the Folio preserves is a revised version of the play, undertaken to make the 1611 text appropriate to the events of 1613.

These inferences have so conditioned recent views of the play that it will be as well to deal with them at the outset. A record of performance at court implies neither a play written specifically for the court nor a first performance there. Dryden, completing Davenant's revision of *The Tempest* for the Restoration stage,

[1] See below, pp. 58–9.

[2] Chambers, *William Shakespeare*, ii. 342. For full references for works cited repeatedly in the commentary and introduction, see pp. 90–2.

[3] Ibid., p. 343.

describes it as having been a Blackfriars play :[1] this may or may not be correct, but there is nothing in the evidence of the Revels Accounts to contradict it. The fact of a court performance need not even indicate that the play was new to the company's repertoire in that season, though we can say on the basis of other evidence that this was the case with *The Tempest*. That a play was presented at court on a particular occasion may indicate that it was chosen for its appropriateness, or that it was revised to suit the occasion; or it may indicate nothing of the sort: numerous examples exist of all three possibilities. *The Tempest* can be shown to have strong affinities with Hallowmas, the occasion of its first recorded court performance,[2] but if we wish to argue its special relevance to Princess Elizabeth's wedding we must deal with the thirteen other plays presented along with it: the list includes such seemingly ominous items as *Othello* and *The Maid's Tragedy*, among a miscellany defying easy categorization.

The only conclusion one can draw from this evidence is that plays were considered appropriate entertainments for weddings – or at least for this one. Even the presence of a masque is no evidence that a courtly venue was intended, and in this case it may, in fact, imply just the reverse: the mechanics of the masque and the apparitions in *The Tempest* are those of the public theatre —descents and ascents, properties appearing and disappearing through trapdoors—not of the Banqueting House, with its changeable scenery. Indeed, if one accepts C. Walter Hodges's argument in *The Globe Restored*, public theatres may well have had flying machinery;[3] whereas Inigo Jones had no such devices at court until his stage included a fly gallery in the 1630s. If, therefore, we wish to think of Ariel as entering flying at 3.3.52.2, he may have done so at the Blackfriars or the Globe, but not at court.

Ceres' allusion in the masque to 'this short-grassed green' (4.1.83) has been taken as evidence that the surviving text of the play was specifically intended for production in the Banqueting House, which was carpeted with green cloth during the performance of masques. But the allusion is just as likely to refer to the

---

[1] Preface, *The Tempest or the Enchanted Island* (1670), in *Works of John Dryden* (1956–), vol.10 (*Plays*), eds. Maximillian E. Novak and George Robert Guffey (Berkeley and Los Angeles, 1970), p. 3.

[2] See John B. Bender, 'The day of *The Tempest*', *ELH*, XLVII, 2 (1980), 235–58.

[3] 2nd edn. (1968), p. 21.

Blackfriars, where the stage was covered with fresh green rushes, and may also have been appropriate to a production at the Globe.[1] That the first court performance did take place in the Banqueting House, rather than in the Cockpit-in-Court (the palace theatre) seems clear enough; and, following Ernest Law's Shakespeare Association Pamphlet *Shakespeare's 'Tempest' as Originally Produced at Court* (1920), the fact has often been taken to imply, once again, that the play was written for the court and produced like a masque. Once again, the evidence will not support these inferences.

The Revels Accounts for 1611 tell us only that *The Tempest* was performed 'at Whitehall'. The information that it was staged in the Banqueting House, rather than in the Cockpit, derives from an entry in the Audit Office accounts for October 1611: 'To James Maxwell gentleman usher . . . for making ready . . . the Banqueting House there [at Whitehall] three severall tymes for playes . . .'[2] Since the King's Men were paid for performing *The Tempest*, *The Winter's Tale*, and one other play at court in late October and early November, the conclusion is reasonable, though not inescapable, that these were the plays for which Maxwell was preparing the Banqueting House during October. But this information becomes less significant the farther one pursues it. Not only *The Tempest*, but all three plays were performed in the Banqueting House; and in December, Maxwell was paid for preparing the same hall for six more plays.[3] Clearly the Banqueting House was not selected for its special appropriateness to *The Tempest*. The fact of a performance there tells us no more about this play, or the conditions of its production, than it tells us about any of the others. We have a precisely analogous case in *Othello*, the first recorded performance of which took place in the Banqueting House in 1604: this does not imply that *Othello* was thought to be particularly masque-like. Indeed, there was not even any necessary association between the Banqueting House and theatrical entertainments, to say nothing of masques. In the next March and April, Maxwell was making the hall ready 'twoo severall tymes for dauncing and another tyme for

---

[1] See W. J. Lawrence, 'The Evolution of the Tragic Carpet', in his *Those Nut-Cracking Elizabethans* (London, 1935), pp. 97–114, and Irwin Smith, *Shakespeare's Blackfriars Playhouse* (New York, 1964), p. 318.

[2] Audit Office, Declared Accounts, Bundle 389, Roll 49, fol. 10b (in the Public Record Office).

[3] Fol. 11a.

bearbating'.[1] The Banqueting House became a place of high decorum in later years, but at this period it was simply one of a number of locations at court available for the presentation of entertainments of all kinds.

This is not to say that the fact of royal patronage and a court audience did not have important effects on the King's Men and on the Jacobean Shakespeare.[2] But they are effects that can be discerned in particular plays only in rare and special instances (a case can be made, for example, for the surviving version of *Macbeth*) and there is no reason to believe that *The Tempest* constitutes one of these instances. The figure of Prospero may well have had something to do with King James in Shakespeare's mind, but if we wish to account for it by invoking the company's royal patron, we will have to explain why the King's Men did not continue to commission and produce plays about magician–monarchs until the end of the reign in 1625.

The Genre. Generic considerations have also had significant effects on attitudes towards the play in this century. Modern criticism has removed *The Tempest* from its place as the first of the comedies, and has invented for it, *The Winter's Tale*, *Cymbeline*, and *Pericles* the category of romance.[3] Modern conceptions of genre are not those of the Renaissance, and our categories tend towards different ends : ours are exclusive and definitive, theirs tended to be inclusive and analytic. To find a new category for a play was not, for the Renaissance critic, to abandon the old ones. J. C. Scaliger describes the *Oresteia* as both a tragedy and a comedy; analogously, the Quarto of *Troilus and Cressida* declares it witty 'as the best comedy in Terence or Plautus', while the Folio editors included the play among the tragedies. These claims do not contradict each other.

---

[1] Fol. 11a.

[2] See, for example, the interesting suggestions in David Bergeron's *Shakespeare's Romances and the Royal Family* (Lawrence, 1985), especially chap. 4.

[3] Coleridge referred to *The Tempest* as a romance in *Notes on 'The Tempest'*, but the term was first used to define a category of Shakespearian drama by Dowden: 'There is a romantic element about these plays. In all there is the same romantic incident of lost children recovered by those to whom they are dear—the daughters of Pericles and Leontes, the sons of Cymbeline and Alonso. In all there is a beautiful romantic background of sea or mountain. The dramas have a grave beauty, a sweet serenity, which seem to render the name 'comedies' inappropriate; we may smile tenderly, but we never laugh loudly, as we read them. Let us, then, name this group consisting of four plays, Romances.' (*Shakespeare* (New York, 1877), pp. 55–6.)

We have invented the category of romance because we believe that certain kinds of seriousness are inappropriate to comedy and because we are made uncomfortable by the late plays' commitment to non-realistic modes. We have, thereby, unquestionably, shed light on the relations between *The Tempest* and three other late plays, but we have also thereby obscured *The Tempest*'s relation to the rest of Shakespearian comedy. And in our imposition of exclusiveness on Renaissance concepts of genre, we have obscured the plays' relation to Shakespearian tragedy as well.[1]

The play is, in fact, as much concerned with tragic as with comic themes: the nature of authority and power; the conflicting claims of vengeance and forgiveness, of justice and mercy; the realities of reconciliation and the possibility of regeneration. It opens with a storm scene that recalls *King Lear* both in its natural violence and in the larger issues it raises about the relation of nature to human authority—issues that are succinctly expressed in the Boatswain's question, 'What cares these roarers for the name of king?' (1.1.16–17). In its concern with political legitimacy and the effects of usurpation, the play reconsiders issues that had occupied Shakespeare's mind from the earliest history plays to *Hamlet* and *Macbeth*. The fact that it centres as well on a happy betrothal has tended to obscure for us its insistent concern with the dangerous potential of sexuality and the uncertain future that marriage represents—themes that recall the examples of Romeo and Juliet, Hero and Claudio, Angelo and Isabella, the worlds of *Antony and Cleopatra* and *Cymbeline*. The rethinking of old issues is mirrored in the play's action: there is a profoundly retrospective quality to the drama, which is deeply involved in recounting and re-enacting past action, in evoking and educating the memory. If there is a path to reconciliation in the play, it is only through this.

Readings and Interpretations. The generic issues are related to questions of character, because in large measure the play contains and controls its tragic potential through the figure of Prospero—the Boatswain's question is, in the context of the second scene, ironic, as we see the storm under the control of the magician of the island. As the text presents him, Prospero is a complex, erratic, and even contradictory figure, though criticism has not invariably seen him

[1] For a fuller discussion see my 'Shakespeare and the Kinds of Drama', *Critical Inquiry*, VI, 1 (1979), 10–23.

1. William Hogarth, Scene from *The Tempest*, c. 1735–40.

as such. The eighteenth century's attitude was, for the most part, announced by Rowe, for whom the play seemed 'as perfect in its kind as almost anything we have' of Shakespeare's, and Prospero's magic had 'something in it very solemn and poetical'.[1] Charles Gildon, in 1710, saw Prospero as almost too serene and untroubled. He conceded that Prospero's account of his past to Miranda 'may seem a little too calm, and that it had been more Dramatic had it been told in a Passion; but if we consider . . . the Story as Prospero tells it, [it] is not without a *Pathos*'.[2] The rage and tension in these speeches are quite evident, but, to an age in search of perfection in Shakespeare, they had become invisible. Hogarth's extraordinary painting of *c.*1735–40 (Fig. 1) implies a similarly sentimental reading. Prospero, looking like a Rembrandt rabbi, watches benignly as a courtly Ferdinand in ermine and gold embroidery salutes a classically draped Miranda, a magic book at her feet and a garlanded lamb at her side. As Robin Simon points out, the iconography derives from conventional annunciation and nativity scenes. Ariel hovers above, a winged cherub with a lute,

---

[1] *The Works of Mr. William Shakespear*, 6 vols. (London, 1709), i. xxiii.
[2] *Remarks on the Plays of Shakespear*, in the so-called volume 7 of Rowe's edition (1710), p. 262.

2. A notably courtly Ferdinand and Miranda,
William Mattocks and Miss Brown at Covent
Garden, 1776, from *The Universal Magazine*.

and Caliban, bearing logs, is apparently oblivious to the fact that
he is crushing one of a pair of linked doves beneath his webbed foot.
Despite his obvious animal qualities and an expression that seems
to combine voyeurism and malevolence, he is obviously part of the
family. The picture is unlikely to represent a stage production, since
until the middle of the century the play was invariably performed
in Davenant's and Shadwell's version, in which Prospero's
household is a good deal larger. But Hogarth's realization is a clear
index to the way his age saw the play.[1] (See Fig. 2.) Henry Fuseli,
at the end of the century, found the model for his Prospero in
portraits of Leonardo da Vinci (Figs. 3–5): the magic had become
both art and science. Hazlitt, writing in 1817, saw Prospero as a
'stately magician', but added that 'the preternatural part has the
air of reality, and almost haunts the imagination with a sense of

[1] Simon also observes that Caliban's appearance derives entirely from Shakespeare's text, not from the current stage tradition. See 'Hogarth's Shakespeare',
Apollo, 109 (March 1979), p. 218. Ronald Paulson sees autobiographical implications in the choice of subject and the depiction of Ferdinand: *Book and Painting*
(Knoxville, 1982), pp. 48–53.

3a (*above*). Henry Fuseli, *The Tempest* 1.2, engraved from the painting for
Boydell's Shakespeare Gallery, 1789
3b (*below*). Detail of Prospero

VITA DI LIONARDO DA VINCI
PITTORE, ET SCVLTORE

4. Giorgio Vasari, *Vite*, Florence 1550, frontispiece to the *Life of Leonardo*.

5. Leonardo da Vinci, so-called self-portrait.

truth'.[1] It was Thomas Campbell in 1838 who first found in Prospero a piece of Shakespearian autobiography, a claim developing out of the assumption that the play was Shakespeare's farewell to the stage. *The Tempest*, he wrote, 'has a sort of sacredness as the last work of the mighty workman. Shakespeare, as if conscious that it would be his last, and as if inspired to typify himself, has made its hero a natural, a dignified, and benevolent magician.'[2] The interdependence of the valedictory and the autobiographical is revealed by the fact that Campbell abandoned the thesis when he was persuaded that *The Tempest* was in fact an early play, dignity and benevolence being apparently, to his mind, unavailable to youthful playwrights. But later critics were not so easily dissuaded, and the notion of Prospero as autobiography has remained solidly within the critical canon.

At the same time—and in certain ways paradoxically—by the mid-century, critics were finding more complexity and a good deal less benevolence in Prospero. Fanny Kemble admirably expresses the ambiguities of feeling that must have been inherent in the acting tradition of the play when she writes of

the noble Prospero, whose villainous policy (not unaided by his own dereliction of his duties as a governor in the pursuit of his own pleasure as a philosopher) triumphs over his fortune, and, through a devilish ability and craft, for a time gets the better of truth and virtue in his person.[3]

Performers, of course, will be more deeply aware than critics of the difficulty of portraying Prospero as utterly benign. For Kemble, the crucial figure in the play, the agent of reconciliation and redemption, was Ariel.

But though many Victorian commentators, like Kemble, found serious problems with the figure of Prospero, the fact did not, in the nineteenth century, alter the benign view of the play as a whole. Historically, there has been a consistent tendency to ignore its ambivalences, sweeten and sentimentalize it, render it altogether neater and more comfortable than the text that has come down to us. There have been exceptions, of course—Lytton Strachey's attack on the play created a small sensation in the early years of

---

[1] *Variorum*, p. 356.
[2] Ibid.
[3] *Variorum*, p. 361.

this century[1]—but none that ever succeeded in presenting a view that turned the play around and became critically normative, in the way that the nineteenth century's sentimental *Cymbeline* has become the complex psychological drama of modern criticism.

The Tempest is a text that looks different in different contexts, and it has been used to support radically differing claims about Shakespeare's allegiances. In recent years we have seen Prospero as a noble ruler and mage, a tyrant and megalomaniac, a necromancer, a Neoplatonic scientist, a colonial imperialist, a civilizer. Similarly, Caliban has been an ineducable brute, a sensitive savage, a European wild man, a New-World native, ugly, attractive, tragic, pathetic, comic, frightening, the rightful owner of the island, a natural slave. The question of correctness is not the issue in these readings; the play will provide at least some evidence for all of them, and its critical history is a good index to the ambivalences and ambiguities of the text. Historical arguments claiming to demonstrate that Shakespeare could not have intended Prospero to be seen as unattractive or Caliban as sympathetic are denying to the Renaissance's greatest playwright precisely that complexity of sensibility which is what we have come to value most in Shakespearian drama, and in Renaissance culture as a whole.

More even than *Hamlet*, the play tempts us to fill in its blanks, to create a history that will account for its action, and most of all for its hero. Indeed, recent psychoanalytic criticism has found in Prospero and in the conflicting wills around him a complex case history—though whose case history is a question that is dealt with less persuasively.[2] Such readings probably testify, again, more to the play's ambivalences and ambiguities than to its psychological consistency. It offers to the psychologizing imagination primarily a world of possibilities, enormously suggestive, but incompletely realized and in significant respects unresolved. It is, indeed, in the

[1] 'Shakespeare's Final Period', in *Books and Characters* (New York, 1922), pp. 47–64.

[2] See, for example, the essays by David Sundelson, Richard P. Wheeler, Coppelia Kahn, and David Willbern in *Representing Shakespeare*, eds. Murray Schwartz and Coppelia Kahn (Baltimore, 1980). Sundelson's essay also appears as the last chapter of his book *Shakespeare's Restorations of the Father* (New Brunswick, 1983), all of which is relevant. The most radical and provocative psychoanalytic study of the play I have seen, and in many respects the most interesting, is Joseph E. Westlund's 'Narcissism and Reparation in *The Tempest*', as yet unpublished, but see his brief discussion of the subject, 'Omnipotence and Reparation in Prospero's Epilogue', in *Narcissism and the Text*, ed. Lynn Layton and Barbara Shapiro (New York, 1986).

matter of the play's resolution that psychological readings, and Freudian readings in particular, tend to have most difficulty accommodating themselves to the text. This is not surprising: if the critical model is a psychoanalytic one, the working out of problems will inevitably be its subject; and if the play is conceived as therapy, the analyst–critic will have a stake in its successful conclusion. In this respect, even very sophisticated Freudian readings tend to be romance readings, ultimately sentimental, emphasizing the promised resolution and ignoring the play's unwillingness to fulfil it.[1]

But all interpretations are essentially arbitrary, and Shakespearian texts are by nature open, offering the director or critic only a range of possibilities. It is performances and interpretations that are closed, in the sense that they select from and limit the possibilities the text offers in the interests of producing a coherent reading. In what follows I have undertaken to be faithful to what I see as the characteristic openness of the text that has come down to us, and to the variety and complexity of its contexts and their implications. To do this is to indicate the range of the play's possibilities; but it is also to acknowledge that many of them (as is the nature of possibilities) are mutually contradictory. There is nothing anomalous in this. The text that has come down to us is poetry and drama of the highest order, but it is also, paradoxically, both less and more than literature. It is, in its inception, a play script written to be realized in performance, with broad areas of ambiguity allowing, and indeed necessitating, a large degree of interpretation. In its own time its only life was in performance, and one way to think of it is as an anthology of performances before Ralph Crane transcribed it for the printer in 1619 or 1620. As a printed text, it is designed to provide in addition the basis for an infinitude of future performances, real and imagined. For all our intuitions of autobiography, the author in it is characteristically unassertive, and offers little guidance in questions of interpretation or coherence. For Shakespeare and his company, the text was only the beginning, not the end, of the play.

In an essay written in 1967, Harry Berger, Jr. epitomized

---

[1] Sundelson, for example, claims that in *The Tempest*, 'the movement of the plot toward fulfillment is the most serene and secure in Shakespeare' (*Representing Shakespeare*, p. 33). For a discussion of some basic conceptual problems with psychoanalytic readings of Shakespeare, see my essay 'Prospero's Wife', *Representations*, 8 (1984), pp. 2–4.

prevailing critical views of *The Tempest* in a quotation from Madeleine Doran: 'The action of the play is Prospero's discovery to his enemies, their discovery of themselves, the lovers' discovery of a world of wonder, Prospero's own discovery of an ethic of forgiveness, and the renunciation of his magical power.'[1] Characterizing this view as sentimental, Berger continues:

I find it hard to accept this reading as it stands, not because it is wrong, but because it does not hit the play where it lives. The renunciation pattern is *there*, but only as a general tendency against which the main thrust of the play strains. There are too many cues and clues, too many quirky details, pointing in other directions, and critics have been able to make renunciation in this simple form the central action only by ignoring those details.[2]

Though Berger's own reading relies too heavily for my taste on allegorical explanations, his assessment of the critical issues seems to me shrewd and accurate, and I have taken his sense of the double and contradictory movement of the play as a whole for my starting-point. The concern with repentance, forgiveness, reconciliation, and regeneration is one that is voiced often throughout *The Tempest*. But a much less clear pattern is the one that is acted out: repentance remains, at the play's end, a largely unachieved goal, forgiveness is ambiguous at best, the clear ideal of reconciliation grows cloudy as the play concludes. And in this respect, the play is entirely characteristic of its author and genre: Shakespearian comedy rarely concludes with that neat and satisfactory resolution we are led to expect, and that criticism so often claims for it. At the end of *Measure for Measure* Isabella gives no reply to the Duke's offer of marriage (and in modern productions she occasionally refuses him); *All's Well that Ends Well* concludes with another test for Helena to pass before she can win Bertram, and the ambiguous observation that 'All yet seems well'; in *Much Ado*

[1] Harry Berger, Jr., 'Miraculous Harp: A Reading of Shakespeare's *Tempest*', *Shakespeare Studies*, 5 (1969), p. 254. The Doran passage is from *Endeavors of Art* (Madison, 1964), pp. 366–7.

[2] Berger's list of sentimental readers includes Reuben Brower, Una Ellis-Fermor, Northrop Frye, and Frank Kermode (p. 280, n. 5); he graciously characterizes an early essay of mine as 'the best defence of this sentimental reading known to me' (p. 254). He observes that 'the hard-nosed view which centers on disapproval of Prospero is much rarer', and cites only Auden's 'The Sea and the Mirror' and Clifford Leech in *Shakespeare's Tragedies* (p. 279, n. 3). A particularly interesting recent essay also belongs in this category: Stephen Miko's 'Tempest', *ELH*, XLIX, 1 (1982), 1–17.

about Nothing Claudio, presented with Hero restored to life, says not a word of apology, gratitude, or love; at the end of *Twelfth Night* we learn that Viola's union with Orsino must be postponed until the missing, enraged Malvolio can be found and persuaded to free the sea-captain who alone can deliver up her clothes; the ladies of France at the conclusion of *Love's Labour's Lost* put off even the discussion of marriage until a year has passed and a penance has been satisfactorily undergone: these are characteristic conclusions of Shakespearian comedy, and that of *The Tempest* is fully consistent with them. How much emphasis we give them, whether we choose to describe them as open-ended, ironic, cynical or realistic, will depend more on our critical predilections than on anything in the texts.

The Tempest opens with a pair of contrasting scenes that may be taken as paradigmatic. The violence and destructiveness of the storm, the helplessness of humanity in the face of nature, are, from the perspective of Prospero and Miranda, an illusion; it is reason and control that are the realities. The fact that reason and control are expressed here as magic, however, complicates matters, and will make the paradigm seem either a great deal more or a great deal less optimistic than its interpretation, depending on whether we think of magic as science or as fantasy. Bacon is often, properly, quoted in support of Prospero's claims, but *Doctor Faustus* and *The Alchemist* are relevant texts too, and the dangers of Prospero's secret studies figure quite as largely in the play as their virtues.[1]

The opening storm scene itself is not at all clear-cut in its implications. Antonio and Sebastian display the brutal natures that they are to exhibit throughout the drama, but where are our sympathies intended to lie in the quarrel between Gonzalo and the Boatswain? Even a Sunday sailor will feel the justice of the Mariners' expostulations, and if we had to predict Gonzalo's character in the rest of the play on the basis of his behaviour here, good sense, cheerful blessedness, and moral uplift would not figure significantly in it. In fact, the themes of insubordination and insurrection are as important in this scene as they are to be in Prospero's

---

[1] David Young writes perceptively of the relationships between *The Tempest* and other magical plays of Shakespeare's age in 'Where the Bee Sucks: A Triangular Study of *Doctor Faustus, The Alchemist*, and *The Tempest*', *Shakespeare's Romances Reconsidered*, eds. Carol M. Kay and Henry E. Jacobs (Lincoln, 1978), pp. 149–66.

account of his history in the scene that follows, and none of the courtiers comes off morally unscathed.

Prospero's long monologue near the beginning of Scene 2 is crucial not only to an understanding of the play's action, but in determining our attitude towards it and towards its protagonist; and here, too, the issues are less clear-cut than they initially appear. A serene sense of power and control, as Prospero calms Miranda's empathetic fears for the ship, is superseded by an account of his past expressed in tortured syntax and punctuated by expletives and retrospective rage. The complexity of the tone is a measure of the complexity of Prospero's, and the play's, relation to the past; this monologue is only the first of a series of repetitions. To narrate his history is to gain control of it, to revise and rectify the past. In doing so, Prospero also creates a past, and thereby a surrogate memory, for Miranda. He will go on to restage his usurpation through the conspiracy of Antonio and Sebastian, to see in Caliban's attempt to kill him a version of his brother's murderous intentions, to find Caliban in Ferdinand and Antonio in Caliban.

The story Prospero tells is a strange mixture of guilt and blame. In it, his usurping younger brother is represented as the villain, but he is also described as acting essentially as Prospero's agent — 'I . . . to him put | The manage of my state' (69–70); 'The government I cast upon my brother' (75)—and Prospero even declares himself responsible for Antonio's dereliction:

'I thus neglecting worldly ends . . . | . . . in my false brother | Awaked an evil nature, and my trust, | . . . did beget of him | A falsehood in its contrary as great | As my trust was' (89–96).

The last assertion is qualified by Prospero's characterization of himself as 'like a good parent,' but this only makes the equivocation more evident. The primary dereliction, in this account, is Prospero's; the abandonment of royal responsibility is the source of much greater evils in the state and in the human condition. The theme is a familiar one in innumerable texts of the period, from *The Book of the Governor* to *Basilicon Doron*, from *Gorboduc* and *Arcadia* to *King Lear*.

Audiences and readers have been much more willing to lay the blame for the disruptions of their societies on Gorboduc, Basilius, and Lear than on Prospero; and Lear's claim that he is more sinned against than sinning, while surely justified, in the context of his drama has scarcely served to mitigate a profound sense of responsibility. And yet *The Tempest*'s ambivalence about Prospero's 'secret studies' is often reiterated, and by Prospero himself. If on the island he has learned to be a good ruler through the exercise of his art (and critics have not invariably been convinced of this), that is also, in Milan, what taught him to be a bad one.

Miranda's role in Prospero's account is both supportive and passive, but the relation between father and daughter is again presented as neither simple nor straightforward, and for directors and performers the scene requires a good many interpretative decisions. Prospero's expostulations take the form of demands for attention and reassurance. Miranda makes it clear that her attention is in no danger of wandering, but her father's violence is retrospective, the playing out of an old rage, and Miranda is not really its object. Prospero's tone towards her is complex and fluctuating, ranging from the gentle dignity of 'I have done nothing but in care of thee, | Of thee, my dear one, . . .' (16–17) to the abrupt and peremptory—'Dost thou attend me?'; 'thou attend'st not!' (78, 87)—to the playful and colloquial: 'Well demanded, wench' (139). (The tone of this last remark was felt to be so problematical that until well into this century it was regularly cut in performance.) Perhaps most startling is Prospero's reply to Miranda's 'Sir, are you not my father?'

> Thy mother was a piece of virtue, and
> She said thou wast my daughter; . . .
>
> (56–7)

This is an old joke, but for interpreters who want Prospero's relation to Miranda to be that of magisterial father to innocent child, it is an inappropriate and disturbing one (Hazlitt's 'stately magician', for example, will have a good deal of trouble with it), and in performance, from the late eighteenth to the early twentieth centuries, it was frequently omitted. Miranda has a curiously parallel moment of her own, responding to Prospero's outraged 'tell me | If this might be a brother' with a defence of Prospero's mother against what she takes to be an implied charge of adultery:

> I should sin
> To think but nobly of my grandmother:
> Good wombs have borne bad sons.
>
> (118–20)

This remark, and the assumptions behind it, cannot be said to spring from Miranda's innocence; and for more than two centuries the performing tradition, with overwhelming uniformity, dealt with what was felt to be an inadmissible contradiction in her character by cutting the entire exchange. It is no doubt true that such moments contribute more to underlying concerns of the play as a whole than to the consistency of its characters, and that any performance will be hard put to do justice to the full range of the text's possibilities, both in its complexity and its genuine inconsistency. But the stage history of the play reveals more than this: a persistent tendency to simplify and flatten Prospero and Miranda, a continuing unwillingness to present them as complex figures at all.

And yet there are moments in this dialogue that imply not only a complex and ambiguous Prospero, but also a very different kind of Miranda from the traditional guileless innocent, one who is conscious of playing a role, conscious of what her relation to her father requires her to say. For example, when Prospero declares his intention of enlightening her about their past she replies that 'More to know | Did never meddle with my thoughts' (21–2); but the claim of passiveness and lack of curiosity is contradicted ten lines later when Miranda says,

> You have often
> Begun to tell me what I am, but stopped,
> And left me to a bootless inquisition,
> Concluding, 'Stay, not yet.'
>
> (33–6)

Directors who decide to underplay the second claim in favour of the first will leave us unprepared for the decidedly active Miranda who indignantly berates Caliban ('Abhorrèd slave . . .', lines 350 ff.), energetically defends Ferdinand against her father's incomprehensible attacks (445 ff.), and disobeys his injunction against speaking with Ferdinand (3.1.36–7, 57–9). Indeed, the passive Miranda was felt by commentators from Dryden and Theobald to the Cambridge editors and Kittredge to require an emended text: 'Abhorrèd slave . . .' was regularly, until well into this century, given to Prospero in editions of *The Tempest*; and even in modern productions, in an age when complexity and ambiguity are common measures of artistic value, the speech is often, still, not Miranda's but Prospero's.

Such examples are enough to indicate the profound gap that has, historically, existed between the textual and the performing traditions of this play. For *The Tempest* to represent the crowning moment of the dramatist's career, to express a sufficiently genial farewell to his art, or even to achieve the reconciliation and restoration that seem to be implied by the genre of comedy, a good deal has always had to be either emended or overlooked.

Wives and Mothers. The drama that Prospero recounts, and that he replays with Miranda, is a family drama; but it is one with a significant absence: the wife and mother. The only reference in the play to Prospero's wife has just been quoted—she 'was a piece of virtue, and | She said thou wast my daughter'. The context implies that, though she was virtuous, women as a class are not, and were it not for her word, Miranda's legitimacy would be in doubt. The legitimacy of Prospero's heir, that is, derives from her mother's word. But that word is all that is required of her in the play; once it has been supplied, Prospero's attention turns to himself and his succession, and he furnishes Miranda with her lineage in a clause that grows increasingly ambiguous:

> . . . thy father
> Was Duke of Milan, and his only heir
> And princess no worse issued.
> (1.2.57–9)

Except for this moment, Prospero's wife is absent from his memory. She is wholly absent from her daughter's memory: Miranda can recall several women who attended her in childhood, but no mother. But the absent presence of the wife and mother constitutes a space that is filled, for Prospero, with surrogates and a ghostly family: the witch Sycorax and her monster child Caliban, who is so often and so disturbingly like the other wicked child, the usurping younger brother Antonio; the good child-wife Miranda, the obedient Ariel, the adolescent and libidinous Ferdinand.

Indeed, Prospero presents his voyage to the island as a way of starting life over again—both his own and Miranda's—and in one extraordinary passage conceives his suffering as a literal childbirth:

> When I have decked the sea with drops full salt,
> Under my burden groaned, which raised in me
> An undergoing stomach to bear up
> Against what should ensue.
> (1.2.155–8)

He has reconceived himself, as Miranda's only parent, but also as the family's favourite child. This has the shape of a Freudian fantasy. He has been banished by his wicked, usurping, possibly illegitimate younger brother Antonio: the younger brother *is* the usurper in the family, and the kingdom he usurps is the mother. On the island Prospero undoes the usurpation, recreating kingdom and family with himself in sole command.

But not quite, because the island is not his alone—or if it is, then he has repeopled it with all the elements of his fantasy, the distressing as well as the gratifying. When he arrives he finds Caliban, child of the witch Sycorax. The figure of Sycorax is largely based on Ovid's account of Medea in Book 7 of the *Metamorphoses*, and indeed the name, which has never been adequately explained, sounds like an epithet for Ovid's witch, the Scythian raven.[1] In Prospero's account, Sycorax was the embodiment of wickedness, but her history is curiously parallel with his. She, too, was a victim of banishment, and the island provided a new life for her, as it did literally for her son, with whom she was pregnant when she arrived. Like Prospero, she made Ariel her servant, and controlled the natural spirits of the island. Sycorax died some time before Prospero's arrival; Prospero never saw her, and everything he knows about her he has learned from Ariel. Nevertheless, she is insistently present in his memory—far more present than his own

[1] The phonetic spelling 'Sythia' was common in the period, and appears twice in the Quarto of *Titus*. The usual derivation of Sycorax's name is from *sus* (pig) and *korax* (raven). I agree with Kermode in finding this improbable, though the second element seems right: ravens are traditional instruments of witchcraft, and are mentioned in connection with the sorcery of both Sycorax and Prospero. Kermode believes that Sycorax is strongly influenced by the Circe legend, and cites 'the mythographic tradition' to the effect that she 'was born in Colchis, in the district of the Coraxi tribe' (p. 26). It is true that Conti says in the *Mythologiae* that she came, like Medea, from Colchis (the Coraxi are not mentioned in the account), and adds that she has sometimes been called Medea's aunt or mother (Frankfurt, 1584, pp. 570, 578). But the Medea story, which immediately follows, and to which Conti's account of Circe is clearly heavily indebted, is more substantial, and more directly relevant to Sycorax. If Circe is involved as a source for the Shakespearian figure, she would provide some basis for the association of pigs with witches: *sus* is what Circe changed her victims into in the *Odyssey*; though in later writers, and throughout the mythographic tradition, they became whatever animal most truly expressed their inner bestial natures. This association, however, seems slight and far-fetched: the *korax* element of the name and the Medea model are both directly accounted for in the text of the play. Finally, it may be relevant to the name of Sycorax's son Caliban that Scythia was traditionally associated with cannibalism, as in *Lear* 1.1.116–17: 'The barbarous Scythian, | Or he that makes his generation messes. . . .'

wife—and she embodies to an extreme degree all the negative assumptions about women that he and Miranda have exchanged.

On the surface, Prospero and Sycorax are antitheses; even posthumously, they are inveterate enemies. But as the play progresses, the similarities between the two sorcerors grow increasingly marked. Sycorax, Prospero says, was possessed by 'unmitigable rage' (1.2.276); she kept Ariel in bondage, and ultimately penned the spirit up in a cloven pine, from which even her own magic was not powerful enough to release him. From Caliban's point of view, and even at times from Ariel's and Ferdinand's, Prospero looks very much like Sycorax. The rage, the demand for unwilling servitude, the continual threats of constriction and painful imprisonment are characteristic of both. And late in the play, the identification of the two in Shakespeare's mind becomes strikingly manifest: Prospero, celebrating and renouncing his magic in a great set piece, does so with a speech of Ovid's Medea (5.1.33 ff.).

Magic. The ambiguities of feeling about magic in the play are a very clear index to its openness. From one aspect, Prospero's art is Baconian science and Neoplatonic philosophy, the empirical study of nature leading to the understanding and control of all its forces. In the *Magnalia Naturae*, Bacon promised, as benefits deriving from the new philosophy, the power to raise storms at will, to control the seasons, to accelerate germination and harvest:[1] in this context, both the opening scene and Prospero's masque constitute a scientific fantasy, marvellous, but not at all inconsistent with reason and virtue. Ralegh in the *History of the World* concurred: magic, he wrote, is 'the connection of natural agents . . . wrought by a wise man to the bringing forth of such effects as are wonderful to those that know not their causes'.[2] Such studies are, moreover, the proper pursuit of monarchs: 'Of the sciences which regard nature, it is the glory of God to conceal a thing, but it is the glory of the King to find it out.'[3] Bacon is presumably recommending royal patronage for science here, not royal scientists; but if Prospero had wanted an apologist for his pursuit of his secret studies, he could have found one in King James's Attorney-General.

[1] *Works*, eds. Ellis, Spedding and Heath (London, 1887–92), iii. 167–8.
[2] Chapter 11, section 2 (1614), p. 202. The quotation has been modernized.
[3] *The Great Instauration*, in R. F. Jones, ed., *Essays*, etc. (New York, 1937), p. 251.

Royal science represents one aspect of Prospero's magic. Another is less solemn: it is theatre, illusionism, the unserious delight implied in Prospero's characterization of his masque as 'Some vanity of mine art' (4.1.41). A darker side is expressed by Prospero himself when he blames his philosophical pursuits for his dereliction of duty; in this view, magic is not a source of power but a retreat from it, and the return to Milan and the reassumption of his dukedom necessitate the renunciation of the art. James himself would have concurred: for all his pride in his scholarship, he distrusted studiousness in monarchs—this is to the point if we wish to view Prospero as a version of the King. In the *Basilicon Doron* he warned his son that 'as for the study of the other liberal arts and sciences [other, that is, than history], I would have you reasonably versed in them, but not pressing to be a pass-master in any of them: for that cannot but distract you from the points of your calling'.[1] Prospero's studies were precisely 'the liberal arts' (1.2.73), and Henry Peacham, embodying James's warning in an emblem book based on *Basilicon Doron*, depicted a royal magician (Fig. 6).[2]

The darkest view of magic is summed up in the figure of Sycorax, that ghostly memory so intensely present in the play, the perverse, irrational, violent, malicious, vindictive principle in nature, progenitor of monsters, lover and agent of the devil on earth. The sources of this figure are not only Ovidian; the study of witchcraft was energetic and prestigious in Shakespeare's England, as the royal dialogue on *Demonology* and Jonson's *Masque of Queens* make clear, and Shakespeare's interest in it is amply attested by *Macbeth*.

Attitudes towards magic in the play, then, range from the most positive to the most negative. Recent criticism has tended to emphasize the tradition of the virtuous mage to the exclusion of everything else. Kermode, following the lead of Frances Yates, sees Prospero as 'a theurgist, whose Art is to achieve supremacy over the natural world by holy magic', and contrasts him with Sycorax, 'a goetist who exploited the universal sympathies, but whose power is limited by the fact that she could command, as a rule, only devils and the lowest orders of spirits' (p. xl). But there is nothing whatever in the play implying that Sycorax's ministers are devils, or

---

[1] *Basilicon Doron*, in McIlwain, p. 40. References to *Basilicon Doron* are to this edition. Quotations have been modernized.
[2] British Library MS Royal 12.A.lxvi, fol. 30r.

6. Henry Peacham, 'Inopportuna Studia', emblem of a royal magician-scientist, from ΒΑΣΙΛΙΚΟΝ ΔΩΡΟΝ, a manuscript emblem book based on the treatise by James I. (Ms. Roy. 12.A fol. 30^r, emblem xxxv.).

that the spirits she controls are any lower, or indeed any other, than those 'weak masters' at Prospero's command. Ariel is the unwilling servant of both.[1]

This is not to say that there is no element of the Neoplatonic

[1] A more detailed argument similar to Kermode's is proposed by C. J. Sisson, 'The Magic of Prospero', *Shakespeare Survey* 11 (Cambridge, 1958), pp. 70–7. Margreta de Grazia, in an extremely well-argued essay, emphasizes the similarities between Prospero and Sycorax: 'Not only are their histories similar and their powers interchangeable, but both sorceress and magician are driven by the same passion —anger' ('*The Tempest*: Gratuitous Movement or Action Without Kibes', *Shakespeare Studies*, 14 (1981), p. 255). She concludes, however, that the essential difference between them lies in Prospero's ultimate repudiation of vengeance and renunciation of magic. These strike me as more problematical than they seem to de Grazia. Walter Clyde Curry considers Shakespeare's debt to Neoplatonic magic in 'Sacerdotal Science in Shakespeare's *The Tempest*', in his *Shakespeare's Philosophical Patterns* (Baton Rouge, 1937). Other magical traditions behind Shakespeare's conception of Prospero are discussed by Barbara Mowat, 'Prospero, Agrippa, and Hocus Pocus', *ELR* 11 (1981), 281–303. The best Neoplatonic reading of the play, and the only one to undertake to account for Prospero's unattractive side, is Karol Berger's 'Prospero's Art', *Shakespeare Studies*, 10 (1977), pp. 211–39.

mage in Prospero, but only that the play is not narrowly systematic in the way such a reading suggests. To view the action of *The Tempest* as depicting Prospero's ascent up a Neoplatonic ladder is to ignore the host of ambivalences and qualifications that are continually expressed about both the central figure and his art. If Prospero in his moment of triumph speaks as Medea, then we have no grounds for making easy distinctions between white and black magic, angelic science and diabolical sorcery. The battle between Prospero and Sycorax is Prospero's battle with himself, and by the play's end he has accepted the witch's monstrous offspring as his own: 'this thing of darkness I | Acknowledge mine' (5.1.275–6).

Caliban. The inner conflict is reflected in the dramatic tension between Prospero and Caliban. Prospero is very hard on Caliban, whose unregenerate nature moves him to a rage that is otherwise reserved only for Antonio. In Prospero's account, there is nothing good about Caliban; if the savage is born to servitude, he is not even a good servant. Prospero's attempts to educate and civilize him have only succeeded in corrupting him. He suffers terribly at his master's hands, but he learns nothing from his suffering—not even how to avoid it. When Miranda charges him with ingratitude, he replies,'You taught me language, and my profit on't | Is I know how to curse' (1.2.362–3). The remark succinctly expresses, and retrospectively justifies, Prospero's view of Caliban.

But we hear much more than curses in Caliban's language. He has a rich and sensuous apprehension of nature, and an imaginative power that is second only to Prospero's. As for the quality of his language, he is given one of the great poetic setpieces in the play:

> Be not afeard, the isle is full of noises,
> Sounds, and sweet airs, that give delight and hurt not. . . .
> (3.2.133–41)

Since the play is so thoroughly under Prospero's control, it is important to observe that this is an instance—perhaps the crucial instance—where what Prospero says and what we perceive do not coincide. Caliban represents a striking failure of Prospero's art; not his only failure—his inability to exact even a hint of repentance from his brother is a more profound one—but the only failure he acknowledges in the play.

Caliban also represents a significant counter-claim to Prospero's authority. The island, he asserts, is rightly his, and Prospero is an invader and usurper. Whatever merit the claim has philosophically, it is allowed to have little dramatically: Caliban is not presented as a noble savage, and his immediate attachment to Stephano is sufficient to confirm Prospero's view of him as a natural servant. The freedom he celebrates in his song in Act 2, 'No more dams I'll make for fish,' turns out, in its refrain, to comprise nothing more than getting a new master.

It is, of course, possible to see in this corrupt behaviour the consequences of Caliban's experience with Prospero—the consequences, that is, of nurture, not nature. In this line of argument, the relation of master and servant, European and native, is modelled on the colonial experience.[1] And in fact, if the dramatic action seems to trivialize the question of Caliban's proper status, the philosophical and legal aspects of his claim to the island have a good deal of resonance throughout the play. They bear not only on the question of Caliban's rights but even more significantly on the nature and sources of Prospero's—or of any ruler's—authority. In a significant sense, it is Caliban who legitimates Prospero's rule. When Caliban tells his master that 'I am all the subjects that you have' (1.2.341), he reminds us that authority may claim to derive from heaven, but in practice it depends on the acquiescence, whether willing or compelled, of those who are governed by it. Miranda would like a world without Caliban, but Prospero is aware that

> ... as 'tis,
> We cannot miss him. He does make our fire,
> Fetch in our wood, and serves in offices
> That profit us.
>
> (1.2.310–13)

[1] The most important treatments of the relevance of colonialism to the play are Octave Mannoni's *Psychologie de la colonisation* (1950), published in England as *Prospero and Caliban*; Stephen J. Greenblatt's brilliant 'Learning to Curse', in *First Images of America*, ed. Fredi Chiapelli (Berkeley, 1976), pp. 561–80, and Trevor R. Griffiths's excellent survey of production history ' "This Island's Mine": Caliban and Colonialism', *Yearbook of English Studies*, 13 (1983), 159–80. Leo Marx's 'Shakespeare's American Fable', in his *The Machine in the Garden* (New York, 1964), pp. 34 ff. and Terence Hawkes's *Shakespeare's Talking Animals* (London, 1973), pp. 194 ff., are both perceptive and interesting, and Leslie Fielder in *The Stranger in Shakespeare* (New York, 1972), pp. 199–253, has good remarks on the play as a colonialist myth. (See also the items cited on p. 49, n. 1 below.)

Aristocracies require proletariats; hierarchies need people at the bottom as well as at the top. There is a great deal of physical labour to be done on the island, and, except for the brief hour of Ferdinand's servitude, only Caliban can be made to do it. What Prospero's magic chiefly enables him to do is control his servants.

As for the legitimate sovereignty of the island, Caliban himself complicates an initially simple issue by deriving his claim from inheritance: 'This island's mine by Sycorax my mother' (1.2.331). He need not do this; the claim could derive from the mere fact of prior possession: he was there first. This, after all, would have been the basis of Sycorax's claim were she to have made one, but it is an argument that Caliban never uses. And in deriving his authority from his mother, he delivers himself into Prospero's hands; for if it is true, as Prospero says, that Caliban is a bastard, 'got by the devil himself | Upon thy wicked dam' (1.2.319-20), then the claim from inheritance is invalid: his illegitimacy would bar him from the succession.

But *is* it true that Caliban is Sycorax's bastard by Satan? Or is this merely more of Prospero's invective, an extreme instance of his characteristic assumptions about women? Nothing in the text will answer this question for us; and it is worth pausing to observe both that Caliban's claim seems to have been designed so that Prospero can disallow it, and that we have no way of distinguishing the facts about Caliban from Prospero's invective about him. There is a paradigm in this for the play as a whole: its realities throughout are largely the products of Prospero's imagination, or of the imaginative recreation of his memory. 'Facts' have a tendency to appear and disappear as Antonio's son does,[1] in a way that defeats any attempt to find in the play a firm history or a consistent world.

Caliban's name seems to be related to Carib, 'a fierce nation of the West Indies, who are recorded to have been *anthropophagi*' (*OED*), from which 'cannibal' derives; and Caliban may be intended as an anagram of cannibal. The implicit assumptions in the choice of the name are clear enough; and they are Prospero's assumptions. But criticism has generally seen much more in Caliban than Prospero does—Joseph Warton observed kindness in his character,[2] and to Coleridge he was 'in some respects a noble

---

[1] See 1.2.439.
[2] *The Adventurer*, no. 97 (9 October 1753).

being: the poet has raised him far above contempt'.[1] Kermode, noting that the Folio text describes Caliban in the Persons of the Play as 'a salvage and deformed slave', relates him to 'the wild or salvage man of Europe, formerly the most familiar image of mankind without the ordination of civility'.[2] The linguistic part of this claim is dubious: 'salvage' has the required sense only in the expression 'salvage man', which appears nowhere in the text. As it is used in the play, and throughout Shakespeare's works, 'salvage' is simply a variant spelling of 'savage'; and it is the form usually used by Ralph Crane, the scribe who prepared *The Tempest* for the press. But the view of Caliban as a familiar European wild man or wodewose is symptomatic of a widespread critical attempt that is prompted by the play itself, to humanize and domesticate Caliban, to rescue him from Prospero's view of him—to succeed with him where Prospero has failed. Auden was responding to the same impulse when, in *The Sea and the Mirror*, he made Caliban the embodiment of suffering humanity. The stage tradition has presented him more often as clownish than frightening, thereby implicitly undercutting the seriousness of Prospero's fears and invective. Beerbohm Tree's famous 1904 production, in which Tree played Caliban, portrayed him as a sensitive and potentially noble creature (Fig. 7), and concluded the play with him at the centre of a pathetic tableau:

Caliban creeps from his cave, and watches the departing ship bearing away the freight of humanity which for a brief spell has gladdened and saddened his island home, and taught him to 'seek for grace'. For the last time Ariel appears, singing the song of the bee . . .—Ariel is now free as a bird. Caliban listens for the last time to the sweet air, then turns sadly in the direction of the departing ship. The play is ended. As the curtain rises again, the ship is seen on the horizon, Caliban stretching his arms towards it in mute despair. The night falls, and Caliban is left on the lonely rock. He is a King once more.[3]

Caliban has generally been seen as a foil to Ariel—the airy spirit, the earthy monster—and Prospero confirms his servant's place in an elemental hierarchy by referring to him as 'earth' (1.2.314). Both long for freedom, and, while only Ariel is offered the hope of

---

[1] *Coleridge's Shakespearean Criticism*, ed. T. M. Raysor (2 vols. 1930), ii 178.
[2] pp. xxxvii–xxxix.
[3] *The Tempest, as Arranged for the Stage by Herbert Beerbohm Tree* (London, 1904), p. 63.

7. Sir Herbert Beerbohm Tree as Caliban, 1904, charcoal drawing by Charles A. Buchel, published in *The Tatler*

obtaining it, in fact both Prospero's servants receive it at the same time, when Prospero resumes his dukedom at the play's end. In contrast to Caliban's elemental sameness, Ariel is volatile and metamorphic. He is male,[1] the asexual boy to Caliban's libidinous man, but (in keeping with his status as a boy actor) all the roles he plays at Prospero's command are female: sea nymph, harpy, Ceres. Though his relation to his master includes a good deal of obvious affection, he is no more a willing servant than Caliban, and Prospero keeps him in bondage only by a mixture of promises, threats, and appeals to his gratitude.[2]

[1] As 3.3.52.2 indicates: '*Enter Ariel, like a harpy, claps his wings . . .*'.
[2] The complexity of their relationship was superbly expressed in Giorgio Strehler's great production for the Piccolo Teatro di Milano, first presented in 1948, and revised in 1977. Ariel, played by a ballerina in a transparent costume over a white body stocking, flies and hovers throughout much of the play; she is in a harness attached to a cable. To free her at the play's conclusion, Prospero undoes the harness, and Ariel with an expression of wonder and gratitude hesitantly takes her first steps into the freedom of the audience. The ability to fly is thus shown to exemplify not Ariel's essential freedom but the spirit's subjection to Prospero.

27

In the dramatic structure of the play, Caliban is even more significantly contrasted with Miranda.[1] The two children have been educated together on the island, the objects of Prospero's devoted care; Miranda has developed into a wonder of civilized grace, Caliban into a surly, malicious and—what is most upsetting to Prospero—a lustful monster. Prospero concludes that Caliban, therefore, is monstrous by nature; but once again the issue is complicated by Prospero's and Miranda's claim that they have taught Caliban everything he knows, and by the clear parallel in Prospero's mind between Caliban and the other wicked child for whose education he claims responsibility, his younger brother Antonio. Prospero acknowledges (indeed, insists on) his complicity in the events that led to his overthrow in Milan. The gesture is symbolically repeated in his final acceptance of Caliban as his own.

Suitors and Rapists. Caliban, like Sycorax, does in fact embody a whole range of qualities that we see in Prospero, but that he consistently denies in himself: rage, passion, vindictiveness; perhaps deepest and most disruptive, sexuality. Theatrically and critically, the most troublesome aspects of the magician's character have been those relating to libidinous energy. Prospero's charge of ingratitude against Caliban, and Miranda's startling denunciation of him, are provoked by the recollection of an attempt by Caliban to rape Miranda. Caliban compounds the offence by acting both unrepentant and retrospectively lecherous:

> O ho, O ho! Would't had been done!
> Thou didst prevent me—I had peopled else
> This isle with Calibans.

> (1.2.348–50)

Caliban's, however, is not the only dangerous sexuality to be feared in the play. Prospero's repeated warnings to Ferdinand against pre-marital sex are not prompted by anything we see of Ferdinand's behaviour. Caliban, in this context, is any man who takes an interest in Miranda, even the suitor of Prospero's choice.

---

[1] Caliban's age can be calculated. Miranda is not yet fifteen—she and Prospero have been on the island for twelve years, and at the time of their expulsion from Milan she was 'not out three years old'. Since Sycorax was pregnant when she came to the island, and died before Prospero arrived, and imprisoned Ariel in the cloven pine a dozen years before that, Caliban is more than ten years older than Miranda, or at least twenty-four.

The ambivalence towards Ferdinand is expressed, too, in the tasks
Prospero sets for him (3.1.9–11), which are, explicitly, Caliban's
tasks. Prospero later apologizes for 'too austerely' punishing Ferdinand (4.1.1), but leaves the young man's offence unexplained.
Nor is Caliban the only criminal Prospero sees in his prospective
son-in-law:

> . . . Thou dost here usurp
> The name thou ow'st not, and hast put thyself
> Upon this island as a spy, to win it
> From me, the lord on't. . . .
> Speak not you for him: he's a traitor.
>
> (1.2.454–61)

The crimes Prospero charges Ferdinand with in this strange moment are those of his brother Antonio: usurpation and treason.

And in fact, Prospero *has* installed another usurper in his kingdom. To provide a husband for Miranda is to acknowledge his own
age and declining powers. Reconciliation and restoration involve
yet another withdrawal from the world of action, in which 'Every
third thought shall be my grave' (5.1.311).[1]

The resolutions Prospero undertakes to provide for the play's
tragic tensions are the traditional ones of comedy: forgiveness of
injuries, repentance for wrongs, above all reconciliation through
the harmony of marriage. He is unambiguously successful in
producing only the third of these; but even here, the promised end
is neither easy to come by nor the union of innocents we have been
led to expect. He interrupts his celebratory masque with the
recollection of another conspiracy against his throne and his life;
and, as the play draws to its conclusion, Ferdinand and Miranda
play out, at chess, a brief game of love and war that seems to
foretell in their lives all the ambition, duplicity and cynicism of
their elders:

MIRANDA
    Sweet lord, you play me false.
FERDINAND   No, my dearest love,
    I would not for the world.
MIRANDA
    Yes, for a score of kingdoms you should wrangle,
    And I would call it fair play.

> (5.1.172–5)

---

[1] For a discussion of Prospero's age, see below, pp. 79–82.

Miranda in this exchange (most commentators to the contrary notwithstanding) is certainly accusing Ferdinand of cheating, and is declaring her perfect complicity in the act. Italian *Realpolitik* is already established in the next generation.

The Renaissance Political Context

Political Marriages. So far we have concentrated on *The Tempest* as Prospero's personal drama. But the play's scope is broader than this; the shipwreck engineered by Prospero brings to the island, both explicitly and by implication, the world of Renaissance politics. In the context provided by Alonso, Gonzalo, Antonio, and Sebastian, family quarrels are public policies, private motives become matters of state. These Italian rulers are returning from the wedding of Alonso's daughter Claribel to the King of Tunis—not a happy occasion, to which the bride went unwillingly, and of which much of the court disapproved.[1] That wedding in the play's background is, on the face of it, very different from the happy union of Ferdinand and Miranda; but both are affairs of state, and the account we are given of Claribel's fate comes immediately after Miranda's first meeting with the man her father has chosen as her husband. If it does nothing else, Claribel's marriage will give us notice that more is at stake in the match Prospero is arranging than the happiness of two young people.

The betrothal of Ferdinand and Miranda is an essential element of the broader reconciliation towards which the play strives. The disarming of traditional enemies through marriage in the next generation is not only a comic convention; it is a piece of Renaissance statecraft. Commentators who see the play mirroring the wedding of Princess Elizabeth to the Elector Palatine are casting their nets too narrowly. Throughout Shakespeare's lifetime the *politique* marriage seemed the likeliest means of resolving the European power struggles, whether, as in France, through the union of Huguenot and Catholic, or, as in the marriages of Mary Tudor to Philip II of Spain, and of James's daughter to the Protestant Prince Frederick, through a consolidation of power. Marriages of reconciliation, of the sort represented by that of Ferdinand and

---

[1] Nothing is said about their reasons for disapproving, but since the King of Tunis would have been a Muslim, the marriage would have been unthinkable to a Renaissance Christian audience.

Miranda, were, however, not popular in England, as the public outrage over Queen Elizabeth's proposed match with the Catholic Duc d'Alençon made abundantly clear. Nevertheless, King James had similar plans for both Prince Henry and Princess Elizabeth: their hands were in his gift; and, despite his assertion in *Basilicon Doron* that 'I would rathest have you to marry one that were fully of your own religion',[1] James was quite clear about the fact that to marry his children to Protestants would have been a waste of good diplomatic currency. By 1611, matches for Henry had been actively negotiated with the Infanta of Spain and a Medici princess (the latter project had been vetoed by the Pope because it did not involve Henry's conversion—that is, precisely because it was ecumenical). At the time of the Prince's death in 1612, James was negotiating for both the French Princess Christine and a Princess of Savoy; Princess Elizabeth's marriage to the Prince of Piedmont was to be part of the latter bargain. Henry, unlike his father, was a militant Protestant, hence his distaste for and distrust of an ecumenical alliance, and his support of the Palsgrave's suit for his sister. Nevertheless, regarding his own marital future, he assured his father of his conviction that it was 'for your Majesty to resolve what course is most convenient to be taken by the rules of State', merely observing that 'my part to play, which is to be in love with any of them, is not yet at hand'.[2] There are no apparent religious overtones to the marriage of Ferdinand and Miranda,[3] but, in so far as it is designed to resolve 'inveterate' territorial enmities (1.2.121–2), it has more in common with James's plans for his children than with the actual wedding *The Tempest* was called upon to help celebrate.

Utopia and the New World. With the shipwreck, more is suddenly at stake on the island, too. Gonzalo's Utopian fantasy (2.1.145–62) brings into the play a whole range of Renaissance thought about the relation of Europeans to newly discovered lands and to their

---

[1] McIlwain, p. 35.

[2] Letter of 5 October 1612, quoted by J. W. Williamson, *The Myth of the Conqueror* (New York, 1978), pp. 138–9.

[3] Though these may be implied: both Milan and Naples were Spanish dependencies during Shakespeare's lifetime, but in the first half of the sixteenth century Naples, under the viceregency of Don Pedro de Toledo, was the centre of a strong Protestant movement which spread throughout Italy until its suppression by the Inquisition.

native populations. These matters would have been especially timely in 1611 because of the recent formation of the Virginia Company, established by royal charter in 1606, which had founded Jamestown in Virginia in 1607, and in 1609, reincorporated as a joint stock company, sent a fleet across the Atlantic bearing four hundred new colonists. The results of this undertaking were little short of disastrous; and at least one episode in the voyage seems to have provided Shakespeare with material for *The Tempest*: during a hurricane near the Virginia coast, the governor's ship was separated from the rest of the fleet, and was driven to Bermuda. The passengers got safely ashore, and wintered comfortably there. William Strachey's account of the adventure is generally considered to have clear echoes in the play. This letter, though not printed until 1625, certainly circulated in manuscript, and Shakespeare was evidently familiar with it. The playwright was associated, moreover, with a number of members of the Virginia Company: Southampton, the dedicatee of *Venus and Adonis* and *The Rape of Lucrece*, Pembroke, who was to be a dedicatee of the Folio, Christopher Brooke, Dudley Digges, and others; and he may have known Strachey. Shakespeare's interest in the venture would have been at least partly personal.[1]

The larger question of the relevance of exploration narratives to *The Tempest*, and of the Virginia Company pamphlets in particular, has been energetically argued since Malone first called attention to them in 1808.[2] Most critics are agreed that at least some of the literature relating to the New World is somewhere behind the play, though E. E. Stoll dismissed the whole matter categorically: 'There is not a word in *The Tempest* about America or Virginia, colonies or colonizing, Indians or tomahawks, maize, mocking-birds, or

---

[1] Selections from the Strachey letter are reprinted in Appendix B. The basic research on Shakespeare and the Virginia Company is in Charles Mills Gayley, *Shakespeare and the Founders of Liberty in America* (New York, 1917), especially chapters 2 and 3. Leslie Hotson, in *I, William Shakespeare* . . . (1937), adds material about Shakespeare and Digges (pp. 203–36).

[2] The best discussion of the significance of travel literature to the play is Charles Frey's '*The Tempest* and the New World', *ShQ* xxx, 1 (1979), 29–41. Philip Brockbank makes some interesting suggestions in '*The Tempest*: Conventions of Art and Empire', in J. R. Brown and Bernard Harris, eds., *Later Shakespeare*, Stratford-upon-Avon Studies, 8 (1966), pp. 183–201. A good summary of material relating to the Virginia expeditions, with some splendid excerpts and some unpersuasive theorizing about Shakespeare's indebtedness to them, will be found in D. G. James, *The Dream of Prospero* (Oxford, 1967), chap. 4.

tobacco. Nothing but the Bermudas, once barely mentioned as faraway places, like Tokio or Mandalay.'[1] Stoll was reacting against claims by Charles Mills Gayley, Sidney Lee, and R. R. Cawley that the play was about the English experience in Virginia; and he was right to insist on their improbability. Stoll's counter-claim that the American experience is utterly irrelevant to *The Tempest* is, however, equally extravagant.

It is not true that 'the still-vexed Bermudas' is the only allusion in the play to the New World. Caliban's god Setebos was a Patagonian deity; the name appears in accounts of Magellan's voyages, and is clear evidence that the Americas were in Shakespeare's mind when he was inventing his islander. The significance of the literature of exploration strikes me as both deeper and less problematic than has generally been argued. Charles Frey's observation that travel narratives provided Shakespeare not with sources but with models, both for the behaviour of New-World natives and for European responses to them, helps to refocus the question and clarify the issues involved.[2]

Certain elements of the play relate to a New-World topos persisting from the earliest accounts until well into the seventeenth century. Hugh Honour, in *The New Golden Land*, reproduces a German broadsheet of 1505 that gives a particularly clear statement of the elements of the topos (Fig. 8).[3] A woodcut shows a group of natives wearing feathers and standing beneath a trellis from which hang human limbs. The text reads, in translation,

The people are naked, handsome, brown, well-built; their heads, necks, arms, genitals, feet of both women and men are lightly covered with feathers. The men also have many precious stones on their faces and chests. No one owns anything, but all things are common property. And the men have as wives those that please them, whether mothers, sisters or friends; they make no distinctions among them. They also fight with each other. They also eat each other, even those who are slain, and hang their flesh in smoke. They live one hundred and fifty years, and have no government.

Cannibalism, Utopia, and free love reappear throughout the century as defining elements of New-World societies. Cannibalism

---

[1] 'Certain Fallacies and Irrelevancies in the Literary Scholarship of the Day', *Studies in Philology*, 24 (1927), p. 487.
[2] '*The Tempest* and the New World', p. 34.
[3] New York, 1975, Fig. 7, p. 12.

8. American natives, woodcut from a German broadsheet, *c.*1505.

especially became part of the standard iconography of America, as Philippe Galle's elegant personification of *c.*1600 makes clear (Fig. 9).[1] If Shakespeare were looking for accounts of New-World natives, an obvious place for him to turn would be to an essay on cannibals. We know he did in fact turn to Montaigne's, where he found the other elements of the topos as well: that the natives have a Utopian government and sanction adultery. The latter observation is especially relevant, because the practice of free love in the New World is regularly treated not as an instance of the lust of savages, but of their edenic innocence; and it may help to explain why Caliban is not only unrepentant for his attempt on Miranda, but incapable of seeing that there is anything to repent for.

But the age's view of the relation of the new to the old world goes deeper than this: it is historical and typological as well. When Thomas Harriot published his account of his voyage to Virginia, he included as an appendix a set of engravings of the ancient Britons, 'for to show', he explains, 'how that the inhabitants of the Great Britain have been in times past as savage as those of Virginia'.[2] *The*

---

[1] Reproduced in Honour, Fig. 77, p. 87.
[2] *A Briefe and True Report of the New Found Land of Virginia* (1590), sig. E[r].

34

9. Philippe Galle, *America*, c.1600.

True Picture of One Pict (Fig. 10) may be compared with Galle's personification *America*. There is no suggestion in Harriot's accompanying text that the early Britons were cannibals, but the analogy between the cultures is clear from the iconography. In the New World, Europe could see its own past, itself in embryo. This is another reason for Prospero to acknowledge Caliban as his own.

Caliban has almost nothing in common with the prelapsarian savages described in Montaigne's essay 'Of the Cannibals', from which Gonzalo's Utopia is, in Florio's translation, almost verbatim derived. He owes more to concepts of the natural depravity of New-World populations, such as are found in explorers' accounts from Purchas to Captain John Smith.[1] In Montaigne, on the contrary, it is the Europeans who are predatory and savage; Shakespeare, as he so often does, dramatizes both sides of the debate, and

---

[1] Kermode cites, for example, Smith's opinion that the Indians are 'perfidious, inhuman, all Savage' (p. xxxvi). Relevant European attitudes are discussed in Greenblatt, 'Learning to Curse' (see above, p.24, n. 1).

10. Theodore de Bry, *The True Picture of One Pict*, from Thomas Harriot, *A brief and true report of the new found land of Virginia*, Frankfurt, 1590.

in the process renders a resolution to it impossible. Montaigne's point in introducing Plato's ideal republic is that, if philosophers could see savage societies, they would have to abandon their Utopian fantasies: New-World natives have created an ideal community that outdoes Plato's imagined one. But Caliban provides no counter-argument to Gonzalo's fantasy. Shakespeare has taken everything from Montaigne except the point.

Authority. Gonzalo's Utopian fantasy is naive and self-contradictory, as the pragmatists in Alonso's party are quick to observe. Nevertheless, in the context of Gonzalo's commonwealth and the assumption behind it that any new land is there for the taking and refashioning, Caliban's accusations against Prospero of usurpation and enslavement reveal an unexpected solidity. Few Renaissance theorists considered the claims of native populations seriously, and Prospero does not undertake to refute Caliban's charges. He assumes his authority, and rules by virtue of his ability to do so. But precisely for that reason the question of authority —on the island, or in any state—remains open.

What is the nature of Prospero's authority and the source of his power? Why is he Duke of Milan and the legitimate ruler of the island? Power, as Prospero presents it in the play, is not inherited but self-created: it is magic, or 'art', an extension of mental power and self-knowledge, and the authority legitimizing it derives from heaven—'Fortune' and 'Destiny' are the terms used in the play. It is Caliban who derives his claim to the island from inheritance, from his mother.

In the England of 1610, both these positions represent available, and indeed normative ways of conceiving of royal authority. James I's authority derived, he said, both from his mother and from God. But deriving one's legitimacy from Mary Queen of Scots was an ambiguous claim at best, and James always felt exceedingly insecure about it. Elizabeth had had similar problems with the sources of her own authority, and they centred precisely on the question of her legitimacy. To those who believed that her father's divorce from Katherine of Aragon was invalid (that is, to Catholics) Elizabeth had no hereditary claim; and she had, moreover, been declared legally illegitimate after the execution of her mother for adultery and incest. Henry VIII maintained Elizabeth's bastardy to the end; her claim to the throne derived exclusively from her designation in the line of succession, next after Edward and Mary, in her father's will. This ambiguous legacy was the sole source of her authority. Prospero at last acknowledging the bastard Caliban as his own is also expressing the double edge of kingship throughout Shakespeare's lifetime (the ambivalence will not surprise us if we consider the way kings are represented in the history plays). Historically speaking, Caliban's claim to the island is a good one.

Royal power, the play seems to say, is good when it is self-created, bad when it is usurped or inherited from an evil mother. But of course the least problematic case of royal descent is one that is not represented in these paradigms at all, one that derives not from the mother but in the male line from the father: the case of Ferdinand and Alonso, in which the wife and mother is totally absent. If we are thinking about the *derivation* of royal authority, then, the absence of a father from Prospero's memory is a great deal more significant than the disappearance of a wife. This has been dealt with in psychoanalytic terms, whereby Antonio becomes a stand-in for the father, the real usurper of the mother's kingdom;[1] but here again the realities of contemporary kingship seem more enlightening, if not inescapable. James in fact had a double claim to the English throne, and the one through his father, the Earl of Darnley, was in strictly lineal respects somewhat stronger than that of his mother. Both Darnley and Mary were direct descendants of Henry VII, but under Henry VIII's will, which established the line of succession, descendants who were not English-born were specifically excluded. Darnley was born in England, Mary was not. In fact, Darnley's mother went from Scotland to have her baby in England precisely in order to preserve the claim to the throne.

King James rarely mentioned this side of his heritage, for perfectly understandable reasons. His father was even more disreputable than his mother; and, given what was at least the public perception of both their characters, it was all too easy to speculate about whether Darnley was in fact his father.[2] For James, as for Elizabeth, the derivation of authority through paternity was extremely problematic. In practical terms, James's claim to the English throne depended on Elizabeth *naming* him her heir (and here we may recall Miranda's legitimacy depending on her mother's word), and James correctly saw this as a continuation of the protracted negotiations between Elizabeth and his mother. His legitimacy, in both senses, thus derives from two mothers, the chaste Elizabeth and the sensual Mary, whom popular imagery represented respectively as a virgin goddess—'a piece of virtue'—and a lustful and diabolical witch. James's sense of his own place in the kingdom is that of Prospero, rigidly paternalistic, but incorporating the maternal as well: the King describes himself in *Basilicon Doron* as 'a loving nourish father' providing the commonwealth with 'their own nourish-milk'.[3] The very etymology of the word authority confirms the metaphor: *augeo*, increase, nourish, cause to grow. At moments in his public utterances, James sounds like a gloss on

---

[1] Kahn makes this point, following a suggestion of Harry Berger, Jr., in 'The Providential Tempest and the Shakespearean Family', in *Representing Shakespeare*, p. 238. For an alternative view, see the exceptionally interesting discussion by Joel Fineman, 'Fratricide and Cuckoldry: Shakespeare's Doubles', in *Representing Shakespeare*, p. 104.

[2] The charge that he was David Rizzio's child was current in England in the 1580s, spread by rebellious Scottish Presbyterian ministers. James expressed fears that it would injure his chance of succeeding to the English throne, and he never felt entirely free of it.

[3] McIlwain, p. 24.

Prospero: 'I am the husband, and the whole island is my lawful wife; I am the head, and it is my body.'[1] Here the incorporation of the wife has become literal and explicit. James conceives himself as the head of a single-parent family. In the world of *The Tempest*, there are no two-parent families. All the dangers of promiscuity and bastardy are resolved in such a conception—unless, of course, the parent is a woman.

The point here is not that Shakespeare is representing King James as Prospero and/or Caliban, but that these figures embody the predominant modes of conceiving of royal authority in the period. They are Elizabeth's and James's modes too.

Epic and History

Italy and Carthage. Dynastic issues and questions of royal authority have an epic as well as a political dimension in the play. In particular, allusions to and echoes of the *Aeneid* are insistent in *The Tempest*, though few commentators have felt sure of what to make of them. Kermode records his conviction that 'Shakespeare has Virgil in mind' (p. xxxiv, n. 2), but does not pursue the matter, and Geoffrey Bullough found nothing from the *Aeneid* worthy of inclusion in *Narrative and Dramatic Sources of Shakespeare*. The most suggestive studies are those of J. M. Nosworthy, Jan Kott, John Pitcher, and Robert Wiltenburg, all of whom find Virgil's influence pervasive and see the meaning of the play as controlled by Shakespeare's response to the *Aeneid*, with Kott arguing that 'the Virgilian myths are invoked, challenged, and finally rejected'.[2] Wiltenburg gives the most tactful statement of the argument, and concurs that 'the *Aeneid* is the main source of the play in this sense, not the source of the plot . . . but the work to which Shakespeare is responding, the story he is retelling'. There are certainly obvious points of contact between the *Aeneid* and *The Tempest*. Ferdinand

[1] From the 1603 speech to Parliament; McIlwain, p. 272.
[2] J. M. Nosworthy, 'The Narrative Sources of *The Tempest*', *RES* 24 (1948), 281–94; Jan Kott, 'The *Aeneid* and *The Tempest*', *Arion*, NS 3, 4 (1976), 424–51 (the passage quoted is on p. 444); John Pitcher, 'A Theatre of the Future: *The Aeneid* and *The Tempest*', *Essays in Criticism*, xxxiv, 3 (July 1984), pp. 193–215. See also Kott's '*The Tempest*, or Repetition', *Mosaic*, x, 3 (1977), 9–36. Wiltenburg's essay, 'The *Aeneid* in *The Tempest*', is in *Shakespeare Survey* 39 (Cambridge, 1987), 159–68. Colin Still's mystical argument in *The Timeless Theme* (1936) includes some perceptive observations.

reacts to his first sight of Miranda with Aeneas' words on seeing Venus, 'o dea certe'[1]—'Most sure, the goddess . . .' (1.2.422); Ceres welcoming Iris in the masque (4.1.76–8) appropriates a Virgilian description of the goddess;[2] Gonzalo, Antonio, and Sebastian, in an exchange that has proved baffling to editors, invoke 'widow Dido' and 'widower Aeneas' and argue over whether Tunis and Carthage are the same place (2.1.75–86); Ariel and his spirits appearing as harpies at Alonso's banquet are re-enacting a Virgilian episode;[3] and if we have Virgil in mind, several other less explicit parallels will suggest themselves.[4] The heroic, with its overtones of the tragic, is a clear strain in the play, and some of its particularity is provided by Virgilian echoes and allusions. The geographical world in which the play is located is largely that of the *Aeneid*: Alonso's shipwreck interrupts a voyage retracing Aeneas', from Carthage to Naples. Shakespeare's imagination in this was the imagination of Renaissance imperialism as well: if Alonso coveted a Tunisian alliance enough to marry his daughter to a Muslim king, Spain took more militant measures, sending invading warships to Tunis several times during the sixteenth century; and Richard Eden records that the Spanish named a harbour in the West Indies Carthago—the newest empire replicating the oldest.[5]

As for the notorious exchange about widow Dido, it has proved baffling only because editors and critics have limited their attention to Virgil. From antiquity until well into the seventeenth century there were two traditions concerning Dido. In the older, which was considered the historical one, she was a princess of Tyre married to her uncle Sychaeus, a priest of Hercules. Her brother Pygmalion, the tyrant of Tyre, murdered her husband for his enormous wealth, and she fled by ship, taking with her both his gold and a group of discontented noblemen. On Cyprus she collected

---

[1] *Aeneid*, i. 328.

[2] *Aeneid*, iv. 700 ff.

[3] *Aeneid*, iii. 225 ff.

[4] e.g. the storm in Act 1, Scene 1, which, however, in its specific details seems to owe more to both the Strachey letter and Ovid's storm in the *Metamorphoses* 11.474–572—see especially lines 539–43. Ferdinand and Miranda at chess, 'discovered' by Prospero pulling aside a curtain, may involve a recollection of Aeneas and Dido in their cave, in pursuit of less innocent pleasures: compare 'The wand'ring prince and Dido . . . | Curtained with a counsel-keeping cave', *Titus*, 2.3.21–3.

[5] *The Decades of the New World* (London, 1555), pp. 51–2. The name of Setebos also appears in this book.

fifty women—in some versions of the story they were raped and abducted by her followers, to provide, like the Sabine women for the Romans, a breeding stock for the new realm. (The imperial mythology, in which rape is essential to the foundation of empire, offers yet another context for Caliban's designs on Miranda.) Dido then sailed to North Africa, where through a combination of shrewd bargaining and deceptiveness she obtained the land to found Carthage. She was an exemplary ruler, famous for her chastity and her devotion to the memory of her murdered husband. She committed suicide to prevent her forced marriage to a local king.

It is Virgil who introduces Aeneas into the legend, and thereby transforms Dido from a model of heroic chastity to an example of the dangers of erotic passion. Later commentators generally account for the transformation as Virgil's way of explaining the traditional enmity between Carthage and Rome—Hannibal, in this reading, is Dido's revenge. In antiquity both Macrobius and Servius rejected the historicity of the Virgilian story, and the Church Fathers regularly treated Dido as a proto-Christian for her absolute fidelity to her marriage vows. Petrarch in *The Triumph of Chastity* explicitly denies that Aeneas had anything to do with Dido's death, and Boccaccio in his book of heroines, *De claris mulieribus*, rebukes Virgil for lying about her.

George Sandys, in his commentary on Book 14 of the *Metamorphoses*, summarizes the tradition in which Shakespeare was working. Regarding Virgil's Dido, he writes that 'others upon better grounds have determined that this was merely a fiction of Virgil's, and that Aeneas never came thither', and he translates an epigram of Ausonius on Dido which ends,

> So fell unforced; lived undefamed (belied),
> Revenged my husband, built a city, died.[1]

English writers, observing that Dido's given name was Elissa, found in her an easy analogue for Elizabeth; and by 1595, the name in Charles Stephanus' *Dictionarium historicum*, a standard source-book for the age, had become Eliza. 'Dido' is an epithet applied to her after her death, usually explained as meaning 'valiant'.

This is the heroic and moral tradition that Gonzalo is invoking in

---

[1] *Ovid's Metamorphoses Englished* (1632), p. 476.

his comparison of Claribel to 'widow Dido'. The cynical Antonio and Sebastian undercut the allusion by invoking the alternative tradition in which Dido abandons her chastity to an equally unchaste 'widower Aeneas': there are many sympathetic readings of the Virgilian episode in the period, but there was no getting around the fact that Virgil's Dido ends as a fallen woman, conscious of her sin,[1] betrayed and abandoned. Brief as it is, the exchange, with its tiny dialectic of ethical and cynical, encapsulates the play's thematic ambivalence towards human nature and towards the past.[2] It is relevant, too, to Prospero's fears for his daughter's chastity; and, in so far as the play's Virgilian overtones encourage us to see Ferdinand as another Aeneas, Prospero's anxiety will strike us as justified.

Antonio's cynical reading of the Dido story, however, also points away from the classical and heroic and towards Elizabethan assumptions about Renaissance Italy: the Machiavellian and diabolical figure as largely in the world of the play as the noble and philosophical. Shakespeare expresses a similar double vision of Italy in *Cymbeline*. Attempts to find a source for the play's modern history comparable to the *Aeneid* for its classicism have been on the whole unrewarding. In 1868 Halliwell[3] called attention to William Thomas's *Historie of Italie* (1549), where there is an account of Prospero Adorno, a Milanese lieutenant who became Duke of Genoa, allied himself with Ferdinand, King of Naples, and was overthrown and expelled.[4] Recently William Slights has found a version of the story told about another Prospero, surnamed Colonna, in Remigio Nannini's *Civill Considerations upon Many and Sundrie Histories* (London, 1601) which adds the name of Ferdinand's

---

[1] *Aeneid*, iv. 172: 'coniugum vocat; hoc praetexit nomine culpam' ('she calls it a marriage; with that name she covers her sin').

[2] The argument over Carthage and Tunis seems to be making a similar point about the ambiguous relation of present to past. Antonio is technically correct, in that Tunis, though near Carthage, always was a separate town; but after the final destruction of Carthage in AD 698, Tunis took its place as the political and commercial centre of the region, to which, moreover, it gave its name. Gonzalo's assumptions, in describing Tunis as the new Carthage, need not be narrowly topographical, and it is not at all clear that a Jacobean audience would have thought that Antonio was correct here, even about topography. Stephen Batman concludes the entry on Carthage in *Batman Upon Bartholome* with the observation that 'the country where it stood is now called Tunis' (1582, p. 232).

[3] In *Selected Notes on 'The Tempest'*; the passage is quoted in the Variorum, p. 343.

[4] Fol. 181[v].

father—Alphonso—and in a nearby chapter discusses communication with spirits. The chapter makes no reference to Prospero, though Ferdinand's ghost figures significantly in it.[1] If either of these is really a source for the play, it is chiefly valuable as an indication of how little Shakespeare was controlled by the history he was reading; but in any case, such accounts are less significant as the basis for a few names and details than as models for contemporary Italian political behaviour.

The Masque

Jacobean Court Spectacles. Prospero's masque for his daughter's betrothal constitutes the prime example we are shown of his art. As the wedding is a matter of state, the masque is an assertion of princely power: the art that raised the tempest and sent ominous apparitions to its victims—enemies of the state, rebels, invaders, usurpers—is presented now in its ceremonial and celebratory mode.

By 1611, the King's Men would have had a good deal of direct experience with court masques. Since James's accession, Whitehall had seen eleven such productions: eight by Jonson, two by Daniel, and one by Campion. As Gentlemen of the Chamber, members of the royal household, Shakespeare and his colleagues would have been present at these, and may also have performed acting roles in at least some of them.[2] To view Prospero's masque in the context of Whitehall's Christmas festivities requires some caution: the masque in *The Tempest* is not a court masque, it is a

---

[1] pp. 9, 12–14, 16. Slights's discussion appears in 'A Source for *The Tempest* and the context of the *Discorsi*', *Shakespeare Quarterly*, 36 (1985), pp. 68–70 (Prospero Adorno's name is incorrectly given as Ardono). The search for historical figures behind the play in fact offers some possibilities that resonate tantalizingly, but they seem to have more relevance to Caliban than to Prospero: the King of Naples in the Prospero Colonna story is Alphonso V of Aragon (1416–58), who claimed the throne of Naples not by descent from his father but under the will of his mother Joanna II, and asserted his right by conquest. His son Ferdinand, who succeeded him as King of Naples, was illegitimate.

[2] The printed editions of Jonson's masques from *Love Restored* (performed on Twelfth Night 1612) onwards regularly name the King's Men as the performers of the professional parts. No acting company is cited in any earlier masque. The surviving royal household accounts for masques before 1612 record some payments to named musicians, but the actors (who were paid at a much lower rate) are invariably unidentified. See, for example, the records for *Oberon* in *Ben Jonson*, eds. Herford and Simpson, 11 vols. (Oxford, 1925–52) x. 521.

dramatic allusion to one, and it functions in the structure of the drama not as a separable interlude but as an integral part of the action.[1] Nevertheless, it is Shakespeare's most significant essay in this courtly genre, and we must look at it in that light as well.

The Jacobean masque was largely the creation of one poet, and the form as Jonson developed it was both celebratory and educative. It undertakes to lead the court to its ideal self through a combination of satire, exhortation, and praise. The monarch is always its centre, and even in cases where Queen Anne or Prince Henry is the nominal protagonist (as in *The Masque of Queens* and *Oberon*), it is always made clear that the force animating the idealizing vision is the king. Thus at New Year's Day 1611, the court saw, in *Oberon*, a group of satyrs won away from the pleasures of drinking and making love to the virtues embodied in the masque's eponymous hero. The fact that the satyrs' pleasures are the traditional courtly ones is part of Jonson's point: the masque is mimetic as well as celebratory. At its climax, a chorus of fairy knights turned from the prince at their centre to the king at the centre of the audience and sang:

> Melt earth to sea, sea flow to air,
>> And air fly into fire
> Whilst we in tunes to Arthur's chair
>> Bear Oberon's desire
> Than which there nothing can be higher,
> Save James, to whom it flies:
> But he the wonder is of tongues, of ears, of eyes.[2]

The refinement of base matter into ethereal, the resolution of elemental tensions into harmony and wonder come through the

---

[1] Claims such as Dover Wilson's and Irwin Smith's that the masque is an addition intended to make *The Tempest* suitable to the royal wedding are based on pure speculation, and treat the play as a much more significant part of the celebration of that event than in fact it was. They also require us to assume that the text preserved in the Folio is the revision for the 1613 court performance. There is no evidence whatever to indicate that this is the case; and if I am correct about the one visible instance of revision in the text (at 1.2.301–4), it is evidence that suggests just the opposite. It seems to be designed to reduce the number of dancers—not the sort of alteration one would expect if the point was to turn the play into a court celebration. See below, pp. 61–2. The question is effectively disposed of by Kermode, pp. xxii–xxiv. For a cogent discussion of the structural integrity of the masque, see Robert Grudin, 'Prospero's Masque and the Structure of *The Tempest*', *South Atlantic Quarterly*, 71 (1972), pp. 401–9.

[2] ll. 220–6. The text of the Jonson masques used is that of the Yale Ben Jonson, ed. Stephen Orgel (New Haven, 1969).

power of royalty. The movement is characteristically Jonsonian, but its relevance to the idealizing aspects of Shakespeare's vision in *The Tempest* is self-evident. Jonson's masques are always about the resolution of conflict, personified, in this early period of his masque-writing career, in clear symbolic figures: the lustful satyrs and obedient fairies of *Oberon*; the maleficent witches banished by the heroic queens of *The Masque of Queens*; the sphinx of Ignorance set against Love, witty and ingenious in *Love Freed from Ignorance and Folly*; the humours and affections tamed by Reason in *Hymenaei*; or simply, in his two earliest masques, the basic antithesis of Blackness and Beauty.[1] These works also celebrate power in the state. The light of the royal presence turns the Ethiops of *The Masque of Blackness* white; the union of bride and groom in *Hymenaei* mirrors the uniting of the two kingdoms of England and Scotland through the new sovereign; the militant virtue exemplified in the heroines of *The Masque of Queens* is guided and controlled by the pacific virtue of the royal scholar; *Oberon* declares that the heir to the throne owes his power and his place to his father the king.

As the move from conflict to harmony is central to the action of the masque, so antithesis is basic to its structure. Jonson did not produce his first fully imagined antimasque until *The Masque of Queens* in 1609, with its opening scene of witches in a hell elaborately realized by Inigo Jones, but the concept of a structural device embodying whatever opposes or threatens the world of courtly virtue was implicit in the form from the beginning—as is evident from the fact that his first masques were a pair contrasting darkness with light, ugliness with beauty. When Shakespeare allegorizes Prospero's two servants, Ariel and Caliban, as antithetical elements—not only exquisite and monstrous, grateful and ungrateful, but also air and earth—he is thinking in terms that the masque had given to the drama.

The Tempest as a whole has certain obvious qualities in common with the masque as Jonson was developing it. The court audience at Hallowmas 1611 had already seen, two years earlier, infernal sorcery superseded by royal virtue in *The Masque of Queens*; in the

---

[1] *The Masque of Blackness* was performed on Twelfth Night 1605; it was to have been followed and completed the next year by *The Masque of Beauty*. But at the next two Christmas seasons the court required wedding masques of Jonson, and *Beauty* was not presented until 1608. The two were published together, as *The Characters of Two Royal Masques, the One of Blackness, the Other of Beauty*, 1608.

first two months of 1611, they saw ignorance and envy defeated by learning and love in *Love Freed*, and (perhaps most to the point if we are thinking of Caliban and of Prospero's fears for Ferdinand and Miranda) licentiousness curbed and brought to the service of a noble intelligence in *Oberon*. Moreover (if John Orrell's argument is correct), Jonson's wedding masque *Hymenaei* (1606) provided Shakespeare with a structural principle based on the Pythagorean canon of harmonic proportion.[1] Prospero's masque may also owe something to Campion's wedding masque for Lord Hay, presented at Whitehall in 1607, which is set in the Bower of Flora, and is concerned with weighing the claims of virginity against those of marriage, and with reconciling Diana to the loss of one of her nymphs. But the relationship of the Shakespearian masque to the court masques that followed it is more interesting; for the masque of Ceres, Iris, and Juno anticipates important elements of the form in the next decade. If Jonson had an English model for such pastoral masques as *The Golden Age Restored* (1615), *The Vision of Delight* (1617), and *Pan's Anniversary* (1620), works in which royal power is conceived as power over nature and the seasons, it can only have been Prospero's masque—there are Continental analogues, but no other English examples. And Shakespeare's notion that a masque is the projection of a royal vision is something that was to be fully realized only twenty years later, in productions designed for the Caroline court.

That there are ambivalences of feeling about all this in *The Tempest* goes without saying—there are powerful ambivalences in Jonson, too, though they are probably not the same ones. It is clear that Prospero is not the ideal ruler required by the harmonious vision that is so often summoned up in the play. If James, too, was not the unmoved mover that Jonson's vision demanded, the poet could at least believe in the power of philosophy and verse to show the king his true place and what he had to be to occupy it. There was, no doubt, for Jonson, a comfortable and probably a necessary fiction here; for the masque was at least as much the king's form as the poet's, and where we tend to see in it only annual compliments to the monarchy, James would have seen it as just the opposite. It was his own annual celebration of his authority, his assertion of his will, and his realization of his sense of his place in

[1] 'The Musical Canon of Proportion in Jonson's *Hymenaei*', *English Language Notes*, 15 (1978), p. 178, n. 14.

the commonwealth, and in the universe. In Prospero's masque, the vision of a royal dramaturge, Shakespeare is responding to something new in English society, and quintessentially Jacobean.

The Masque as Image and Symbol. Prospero the illusionist, moving his drama towards reconciliation and a new life, presents in the betrothal masque his own version of Gonzalo's Utopia, a vision of orderly nature and bountiful fruition. The performance opens with benign deities, Iris and Ceres. If we miss an antimasque, the drama itself has provided several. Prospero in his fierce preliminary charge to Ferdinand invokes a trio of maleficent personifications attendant on unchaste behaviour:

> If thou dost break her virgin-knot before
> All sanctimonious ceremonies may
> With full and holy rite be ministered,
> No sweet aspersion shall the heavens let fall
> To make this contract grow; but barren hate,
> Sour-eyed disdain, and discord shall bestrew
> The union of your bed with weeds so loathly
> That you shall hate it both.
>
> (4.1.15–22)

Ferdinand's reassuring reply banishes these antimasque figures so that the masque can begin. Recent criticism has judged Ferdinand's protestations to be suspiciously overemphatic here;[1] but his rhetoric functions just as, in *The Masque of Queens*, Jonsonian assertions of heroic virtue had done, to expel vindictive witchcraft and lead in the heroines—and, we may perhaps wish to add, just as unconvincingly.

Two other prime instances of Prospero's art, the opening storm and the harpies' banquet, may also be seen as antimasques to the magician's entertainment, and the figure of Iris is an appropriate exorcist for both. As the rainbow, she embodies, in both the classical and biblical traditions, divine providence. As God's pledge to Noah after the universal flood, she demonstrates that (as George Herbert was to put it) 'Storms are the triumph of his art.'[2] And Iris's connection with the harpies is a family one: they are sisters, daughters of Thaumas, whose name, 'wonder', links him with

---

[1] See, for example, the powerful Freudian reading by David Sundelson in 'So Rare a Wonder'd Father', in *Representing Shakespeare*, p. 48.

[2] 'The Bag', l. 5.

both the thaumaturge Prospero and his daughter Miranda.

Iris, bringing together Ceres and Juno to celebrate the royal betrothal, reconstitutes the fragmented world of the play. As Prospero had incorporated mother and wife in coming to the island, so his art now figures them forth in forms of power and benignity, the primal mother and the wife of Jupiter. Goddess of earth and goddess of air,[1] patronesses of agriculture and of marriage, opposites and complements, together they resolve the dramatic tension implicit in Caliban and Ariel. When earth is seen as Ceres, it is no longer intractable, but productive and nurturing; when air is seen as Juno, it is no longer volatile, but universal and majestic. Even Prospero's libidinous fears are put to rest here: Iris assures Ceres that Venus and Cupid, plotters of the rape of her daughter Proserpine, have fled from the scene, confounded by the chaste vows of Ferdinand and Miranda.

Ceres, say the mythographers, brought civilization to human society. In Abraham Fraunce's summary, she

first found and taught the use of corn and grain, and thereby brought men from that wild and savage wandering in woods and eating of acorns to a civil conversing and more orderly diet, and caused them to inhabit towns, to live sociably, to observe certain laws and institutions; and for these causes was herself made a goddess.[2]

She presides here because Prospero's masque is a civilizing vision, and, in contrast to the bounty of the island, the fertility it invokes is controlled and orderly. Caliban, the provider of Prospero's food, is a hunter and fisherman; the islanders are sustained by wild things, and live as predators on nature. But the masque celebrates agriculture, and refers us to a sophisticated society. Ceres even remarks on 'this short-grassed green' (83): this action takes place on a well-tended lawn—or on the Blackfriars' stage, or the court dancing-floor.[3] The season adduced is, to begin with, 'spongy April', the start of the agricultural year. Crops are not yet sprout-

---

[1] Juno is regularly explained as air in Renaissance mythographies. Vincenzo Cartari succinctly sums up the tradition: 'Quelli li quali dissero che gli antichi sotto il nome di diversi Dei adorarono gli elementi, posero Giunone per l'aria' ('those who say that the ancients worshipped the elements under the names of various gods call Juno the air'), *Le Imagini degli Dei* . . . (Venice, 1571), p. 172. The most detailed and helpful discussion of the mythology of the masque is Yves Peyré's, 'Les "Masques" d'Ariel', *Cahiers Elisabéthains*, 19 (April 1981), 53–71.

[2] *The Countesse of Pembrokes Yvychurch* (1592), pp. 26ᵛ–27ʳ (sigs. G4ᵛ–H1ʳ).

[3] See above, pp. 2–3.

ing; and, similarly, the nymphs are 'cold' and 'chaste', the bachelor with spring fever is 'lass-lorn' (65–8). Within fifty lines, however, Ceres is invoking the full barns and garners of high summer, 'Vines with clust'ring bunches growing, | Plants with goodly burden bowing' (112–13); and shortly before Prospero stops the performance Iris summons 'sunburned sickle-men, of August weary' (134). A whole season of growth, fruition, and harvest has been encompassed in the masque's brief span.

This entertainment, for all its lightness, is re-enacting central concerns of the play as a whole. It invokes a myth in which the crucial act of destruction is the rape of a daughter; it finds in the preservation of virginity the promise of civilization and fecundity, and it presents as its patroness of marriage not Hymen but Juno, the goddess who symbolizes royal power as well. This is Prospero's vision, symbolically expressing how deeply the fears for Miranda's chastity are implicated with his sense of his own power, how critical an element she is in his plans for the future. But she is valuable to him, and an extension of his authority, only so long as she remains a virgin, a potential bride for the husband of his choice. The underlying assumptions here are not unique to Prospero, or to this play, or to Shakespeare. They are implicit, too, for example, in Ralegh's selection of the name Virginia for his new colony. The epithet acknowledges the extent to which virginity had become, in Elizabeth's rein, a crucial attribute of royal power, and had become for James, striving like Prospero and Alonso for political alliances through the marriages of his children, an essential royal bargaining chip. In the potential of virginity lay not only civilization but the promise of infinite bounty within a hegemonic order.[1]

But after autumn, what then? The drama in which Prospero performs is a series of crises, and the pressure of time figures significantly in it: *tempestas*, the tempest of the title, has as its root

[1] Philip Brockbank considers the relevance of the Virginia colony to the masque in '*The Tempest*: Conventions of Art and Empire', in *Later Shakespeare*, pp. 195–6. Kott's '*The Tempest*, or Repetition' (see above, p. 39, n. 2) includes a provocative discussion. The most detailed and interesting treatment to my knowledge is by John Gillies, 'Shakespeare's Virginian Masque', forthcoming in *ELH*. (See also the items cited on p. 24, n. 1, above.) Glynne Wickham's 'Masque and Antimasque in *The Tempest*', *Essays and Studies 1975*, ed. Robert Ellrodt (1975), pp. 1–14, reads the masque in relation to Elizabeth I, the Jacobean royal family, and the European political context; the argument seems to me too narrowly conceived, but the material it presents is important.

tempus. Prospero's masque, in contrast, moves in an easy progression; and as its eros includes no lust, its natural cycle includes no winter. Ceres, indeed, makes this one of her gifts to the lovers: 'Spring come to you at the farthest, | In the very end of harvest' (114–15): after autumn, spring will return at once.

This is the point at which Prospero interrupts his masque, and in so doing suddenly brings to it a recognition of what has been omitted from the vision of the ideal. Missing from the revels are violence, lust, death, and, above all, a sense of the importance of the moment, of time as a series of crises. This sense has filled the play; Prospero's awareness of the drama of time, his ability to seize the instant, in large measure constitutes the source of his power. But his imaginative creation has made him forgetful of the realities of the world of action: everything the masque excludes is now impending in the conspiracy of Caliban, Stephano, and Trinculo.

Critics have tended to underestimate the seriousness of this moment, observing that the plotters are inept and comic, and pose no real threat. The threat, however, is less in the conspiracy itself than in Prospero's forgetfulness of it: this is the first moment when the magician loses his awareness of the play's continuing action, and is in danger thereby of losing his control over it. Just as Caliban is replaying Antonio's conspiracy, so Prospero finds himself once again relinquishing his power to the vanities of his art. Forgetfulness has an added dimension now, too. It is an aspect of the sense of age and infirmity that is, for Prospero, a consequence of Miranda's marriage, and that, after the masque, he so sharply expresses.

Renunciation and Resolution

As Prospero presents it, the resolution of the play depends on his willingness to perform several acts of renunciation, chief among which is the abandonment of his magic. The enchanter's powers are, in large measure, explicitly theatrical. He has at his command a troupe of actors headed by Ariel, and all the resources of the Jacobean stage—flying devices, trapdoors, ascents and descents, appearing and disappearing theatrical properties, even a wardrobe of costumes, the 'glist'ring apparel' that Ariel produces from Prospero's cell, and that proves so fatally tempting to Stephano and Trinculo.

Except for the storm at the play's opening, Prospero's power is exemplified as power over people. It has been customary not only to talk about the magician as a Renaissance scientist, but to see alchemical metaphors in the grand design of the play. No doubt there is something in this; but what the action presents is not experiments and empiric studies but a fantasy about controlling other people's minds. Does the magic work? We are shown a good deal of evidence of it: the masque, the banquet, the harpies, the tempest itself. But the great scheme is not to produce illusions and good weather, it is to bring about repentance and reconciliation; and here we would have to say that it works only indifferently well. 'They being penitent,' Prospero tells Ariel, 'The sole drift of my purpose doth extend | Not a frown further' (5.1.29–30). There is a large concession in this, the renunciation of the rage and vengeance that have determined so much of Prospero's tone throughout the play; but the promised resolution is not in fact to be realized. Prospero's assertion opens with a conditional clause whose conditions are not met. Alonso is penitent, but the chief villain, the usurping younger brother Antonio, remains obdurate.

Why does Prospero declare himself satisfied with this conclusion? There is a strong sense of displacement here, a sense that the demand for repentance has been deflected from Antonio to Alonso; and indeed, throughout the play Prospero's efforts have been much more powerfully directed towards Alonso than towards his brother. Nothing, the action seems to say, not all Prospero's magic, can redeem Antonio from his essential badness; but the corollary to this is that Prospero's magic has not, on the whole, been employed to bring about the reform of Antonio. And since Shakespeare was free to have Antonio repent if that is what he had in mind (half a line would have done for critics craving a reconciliation), we ought to take seriously the possibility that that is not what he had in mind.

Why, to begin with, in returning to Milan, does Prospero renounce his magic? Most commentators explain that he gives up his power when he no longer needs it. This is an obvious answer, but it strikes me as too easy, a comfortable assumption cognate with the view that the play concludes with repentance, reconciliation, and restored harmony. To say that Prospero no longer *needs* magic is to beg all the most important questions. What does it mean to need magic? Did Prospero ever need it, and if so, why?

And though he talks a good deal about renouncing it, does he in fact give it up?

The play's attitudes towards magic are, as we have observed, profoundly ambivalent. Magic allies the enchanter with the forces of nature, with Fortune, Destiny, 'providence divine'; but Prospero's devotion to his secret studies is also the source of all the discord in his past; even, as he presents the matter, the cause of his brother's 'evil nature' (1.2.93). The question of the need for magic goes to the heart of how we interpret Prospero's character: is the magic a strength or a weakness? The answer will be, at best, an ambiguous one. To say that he no longer needs magic is to say that his character changes in some way for the better; that, by relinquishing his special powers, he becomes at last fully human. This is certainly what Prospero proposes for himself at the beginning of Act 5:

> Though with their high wrongs I am struck to th' quick,
> Yet with my nobler reason 'gainst my fury
> Do I take part. The rarer action is
> In virtue than in vengeance. . . .
>   . . . this rough magic
> I here abjure; and when I have required
> Some heavenly music . . .
>   . . . I'll break my staff,
> Bury it certain fathoms in the earth,
> And deeper than did ever plummet sound
> I'll drown my book.
>
>   (5.1.25–57)

To renounce magic is also, here, to renounce vindictiveness and vengeance.

But is this in fact what happens at the end of the play? Let us consider Prospero's renunciation—not only what he says he will do, but what he does. What does it mean for the magician to give up his power? Letting Miranda marry and leaving the island are the obvious answers, but they can hardly be right. Miranda's marriage is brought about by the magic; for all the evident pain of losing his daughter, her betrothal to Ferdinand is part of Prospero's plan. It pleases Miranda, certainly, but it is designed by Prospero as a way of satisfying himself. Claribel's marriage to the King of Tunis looks less sinister in this light: daughters' marriages, in royal families at least, are arranged primarily to please their fathers. And leaving the island, reassuming the dukedom, is part of the plan too. Both of these are presented as acts of renunciation, but they are in fact what the exercise of Prospero's magic is intended to effect, and they represent his triumph.

Prospero promises to renounce his art in the great monologue at the beginning of Act 5, 'Ye elves of hills, brooks, standing lakes, and groves'. For all its valedictory quality, this is the most powerful assertion of his magic the play gives us. It is also, as we have seen, a powerful literary allusion, a close translation of a speech of Ovid's Medea, and it makes at least one claim for Prospero that is made nowhere else in the play: that he can raise the dead. At this moment Prospero is all enchanters, black and white, classic and modern. The claim of omnipotence powerfully expresses the magnitude of what is to be given up, but the play includes no comparable moment for the renunciation itself. In fact, we will look in vain for Prospero breaking his staff and drowning his book. The promise, here, is everything.

If it is an unambiguous promise, the move towards reconciliation is far less so. Alonso is welcomed, embraced, and forgiven in short order (5.1.109–11), but Antonio poses a greater problem. 'You, brother mine, that entertained ambition, | . . . I do forgive thee' (5.1.75–8), Prospero says, and then qualifies the pardon at once ('Unnatural though thou art'), reconsiders it as more crimes are remembered, some to be held in reserve ('At this time | I will tell no tales' (128–9)), all but withdraws it ('most wicked sir, whom to call brother | Would even infect my mouth' (130–1)), and only then confirms it through forcing Antonio to relinquish the dukedom, an act that is presented as something he does unwillingly. It is important to observe that Antonio does not repent here —he is, indeed, not *allowed* to repent. Even his renunciation of the crown is presented as compelled; the act is Prospero's:

> I do . . . require
> My dukedom of thee, which perforce I know
> Thou must restore.
>
> (131–4)

The crime that Prospero holds in reserve for later use against his brother is the attempted assassination of Alonso. The episode takes place at 2.1.183 ff.: Prospero has sent Ariel to put all the shipwreck victims to sleep except Antonio and Sebastian. Antonio then persuades Sebastian to murder Alonso and thereby become King of Naples. Sebastian agrees, on the condition that Antonio kill Gonzalo. At the moment of the murders Ariel reappears and wakes Gonzalo:

> My master through his art foresees the danger
> That you, his friend, are in, and sends me forth—
> For else his project dies—to keep them living.
>
> (2.1.295-7)

This situation has been created by Prospero, and the murderous conspiracy is certainly part of his project—this is why Sebastian and Antonio are not put to sleep. If Antonio is not compelled by Prospero to propose the murder, he is certainly acting as Prospero expects him to do, and as Ariel says Prospero 'through his art foresees' that he will. Prospero is restaging his usurpation, and maintaining his control over it this time.

So, at the play's end, Prospero still has usurpation and attempted murder to hold against his brother, things that still disqualify Antonio from his place in the family. The brothers remain natural and inveterate enemies. Obviously there is more to Prospero's plans than reconciliation and harmony—even in the forthcoming happy marriage of Ferdinand and Miranda. If we look again at that marriage as a political act (the participants are, after all, the children of monarchs), we will observe that, in order to establish the line of succession, Prospero is marrying his daughter to the son of his enemy. There is, as we have seen, good Renaissance statecraft in this. In the play's terms, however, it has curious implications. It has the virtue of excluding Antonio from any future claim on the ducal throne, but it also effectively disposes of the realm as a political entity: if Miranda is the heir to the dukedom, Milan through the marriage will become part of the kingdom of Naples, not the other way round. What this means is that Prospero has recouped his throne from his brother only to deliver it over, upon his death, to the King of Naples once again. The usurping Antonio stands condemned, but the effects of the usurpation, the alliance with Alonso and the reduction of Milan to a Neapolitan fiefdom are, through Miranda's wedding, confirmed and legitimized. Prospero has not really regained his lost dukedom—his 'dukedom yet unbowed': he has usurped his brother's. In this context, Prospero's final assertion that 'Every third thought shall be my grave' (5.1.311) may be something more than the simple acknowledgement of advancing age and diminishing power that will inevitably be its primary effect in performance, a conventional *memento mori*. The remark is also a forecast of victory: Prospero has now arranged matters so that his death will remove Antonio's last link with the ducal power. His grave is the ultimate triumph over his brother. If we look at the marriage in this way, giving away Miranda is as much a means of preserving Prospero's authority as of relinquishing it.

The interpretative issue here is not really why Prospero is incapable of being fully reconciled with his brother. On a much more basic level, it is why Shakespeare, having set up such clear expectations about the matter, was unwilling to have Antonio repent. There may, of course, be a biographical explanation that we can never know; but the play's genre itself points to one kind of answer. Just as Prospero interrupts his masque because the idealizing vision not only misrepresents the reality of his drama but finally threatens it, so Shakespeare, in the development of his comedy, increasingly finds the promised restorations and marriages of comic conclusions inadequate to reconcile the conflicts that comedy has generated. This is not to say that Shakespearian comedy does not end happily, but that its happy ending does not exhaust the energies of the drama.

Neat as its conclusion is, therefore, *The Tempest* in its final moments opens outwards. The lovers have each other, Prospero is the duke again, Alonso is repentant and forgiven, Antonio is, if not defeated, at least at bay, Caliban announces that he will 'seek for grace hereafter', and Ariel is free at last. But Miranda's 'O brave new world | That has such people in't' expresses an irony that is acknowledged even by Prospero in his hour of triumph; and his reply, ''Tis new to thee', implies that there will always be a great deal of unfinished business.

Prospero provides an epilogue that is in one respect unique in Shakespeare's drama. He declares himself not an actor in a play but a character in a fiction; and, instead of stepping out of character, he expands the fiction beyond the limits of the drama. His charms, he says, 'are all o'erthrown'; but he is still able to invest his magical powers in the audience. The spells are now ours; we have become the enabling factor in the fiction. Our breath, not

Ariel's, must send his ship back to Italy, and it is we who must forgive him his faults as a higher power forgives ours.[1] We are, he tells us, his master; but we are also his servant. The sense of unfinished business is finally the life of the play. Prospero's is a story for which Shakespeare provides no ending.

Text and Date

The Text. The earliest text of *The Tempest* is that of the 1623 Folio, where it appears as the first play in the volume. It is a clear text with very few obvious corruptions, carefully and at times idiosyncratically punctuated, and with unusually elaborate stage directions. The latter two points, along with a number of unusual spellings and peculiarities of format, such as the occasional appearance of prose as verse and verse as prose, have led scholars to the conclusion that the printers' copy for the play was a manuscript prepared by the scrivener Ralph Crane. Crane was employed from time to time by the King's Men to make transcripts of plays. A number of these survive, including examples in multiple copies, and one case—Middleton's *A Game at Chess*—in which a manuscript in the author's hand also survives. We, therefore, have a good deal of evidence about Crane's methods and characteristics, though it is evidence that has as yet been only partially evaluated. Crane also prepared the Folio copy for *Two Gentlemen*, *Merry Wives*, *Measure for Measure*, and *The Winter's Tale*. The most thorough study of Crane's role in the preparation of copy for the First Folio is T. H. Howard-Hill's *Ralph Crane and Some Shakespeare First Folio Comedies* (Bibliographical Society of the University of Virginia,

---

[1] Kermode quotes Warburton's observation that Prospero's epilogue 'alludes to the old stories told of the despair of necromancers in their last moments, and of the efficacy of the prayers of their friends for them', and compares a moment late in Marlowe's *Doctor Faustus*: 'FAUSTUS Ay, pray for me, pray for me. . . . SECOND SCHOLAR Pray thou, and we will pray that God may have mercy upon thee.' But he rejects the implications of the analogy: 'Prospero is throughout presented as a theurgist, and here he refers to his renunciation of special powers over the spirit world and his retention of the normal means of access to it. To turn him into a Faustian goetist at this point is to invite confusion.' The confusion, however, is Shakespeare's: ambivalence about the character, and even about the efficacy of the magic, is evident throughout the play, and the articulation of the need for forgiveness and grace concludes a movement that is begun at Prospero's first appearance, in his account of his pursuit of his secret studies and its effects on his character, his family, and his dukedom.

Charlottesville: University of Virginia Press, 1972). Important additional material will be found in Jeanne Addison Roberts's 'Ralph Crane and the Text of *The Tempest*', *Shakespeare Studies* 13 (New York, 1980), pp. 213–33.

Sir Walter Greg, writing in 1955, was summing up two decades of scholarly opinion when he asserted that 'Crane's transcript was clearly made from the author's original'. He continued, 'this must have been carefully prepared and might have served as a prompt-copy. But if the book-keeper annotated it to any extent for the stage, Crane edited away all traces of his handiwork.'[1] Howard-Hill accepts this assumption, though with caution, remarking that the argument involves 'some circularity of reasoning'.[2] But Jeanne Addison Roberts raises important objections:

The evidence is ambiguous and not very substantial. Vague or literary stage directions and variety in speech prefixes, often taken as signs of authorial copy, have recently been shown . . . to occur in extant prompt copies as well. Similarly, stage directions marked early, such as those to be found [in four instances in] *The Tempest* cannot be used with certainty to discriminate between prompt and authorial copy and are indeed common in Crane's manuscripts prepared for reading. . . . All that can be claimed with reasonable confidence is that *The Tempest* does not contain such signs of theatrical use as repeated stage directions and actors' names. The clean-ness of the text would be consistent with that of a copy of author's papers prepared for the printer, or perhaps with an intermediate manuscript between authorial and prompt copy. Because the stage directions are longer and more specific than one would expect of either authorial or prompt copy, I have favored the former hypothesis.[3]

Roberts believes it most likely, that is, that the printer was working from a copy prepared by Crane specifically for publication, and that the manuscript Crane was copying was either Shakespeare's origi-nal or possibly a fair copy of it, and not the playhouse prompt-book that served as the performing text, whether this was a marked-up copy of a Shakespeare holograph or not.

Editorial studies of the play in recent years have focused par-ticularly on the stage directions. These are detailed and circum-stantial, and suggest neither the author's foul papers nor the prompt-book. A strong and convincing case has been made for Crane's sophisticating hand in them. Greg writes:

[1] Greg, p. 420.
[2] *Crane*, pp. 108–9.
[3] Roberts, pp. 213–14.

There can be no doubt that these [stage] directions are fundamentally the author's: they are almost throughout what we have come to expect of Shakespeare in his later plays when he was writing instructions for the producer. But we have not got them, it would seem, exactly as they were written. No author writing with the prompter or the producer in view would use such an expression as 'with a queint device the Banquet vanishes' [3.3.52.3]. It is the language of a spectator recording with appreciation the ingenuity of the staging, and there is no reason why Crane should not have seen *The Tempest*.[1]

Roberts compares the stage directions with those in Crane's transcriptions of *A Game at Chess*, and concludes (though with extreme caution) that they are not inconsistent with accounts of a production.[2] The most thorough examination of the question as a whole is John Jowett's 'New Created Creatures: Ralph Crane and the Stage Directions in *The Tempest*', *Shakespeare Survey 36* (Cambridge 1983), pp. 107–20. Jowett concurs that it is likely that the stage directions involve recollections of a performance, and his analysis of their language in relation to that of other plays of the period has important implications for our sense of the action they are intended to represent, especially in the masque in Act 4, Scene 1. My treatment of the stage directions in this edition is based on Jowett's research. Since preparations for the Folio were under way in 1620, Crane may have been working as early as that year, or perhaps the year before. No contemporary performances of *The Tempest* are recorded other than the two at court in 1611 and 1613 (see below, The Date, p.62), but if the play was in the repertoire of the King's Men, Crane could certainly be recalling a recent production.

The fact that *The Tempest* is placed first in the Folio has figured significantly in a variety of critical speculations, but the bibliographical evidence will lend little support to any of them. Greg observes that

in their arrangement the editors followed the businesslike principle of displaying their most attractive wares in the most prominent positions. Thus the section of Comedies both begins and ends with a series of five plays that had never been printed before, or had only appeared in very inferior versions, and the four of which texts were already available were tucked away between.[3]

[1] Greg, pp. 419–20.
[2] Roberts, pp. 214–18.
[3] Greg, p. 80.

He suggests that the text of *The Tempest* 'may have been prepared specially to serve as a model for the collection'.[1] But Howard-Hill firmly rejects this assumption, and with it, by implication, a complex of critical claims that have been deduced from it. The statement is worth quoting in full:

Observations that *The Tempest* was intended to serve 'as a model for the editing' of the Folio depend on two main arguments, that the text was prepared with some degree of care, and that it was the first play to be printed in the Folio. Neither of these arguments will bear inspection. Such signs of 'editorial care' as the division into acts and scenes, the indication of locality and the list of characters in the play are common in Crane's transcripts, and reveal nothing of his specific intentions for the transcript of *The Tempest*. Nor can it be deduced from these features of the text, or from its accuracy and 'excellent punctuation' that the text itself was given especial attention. The priority of *The Tempest* in the Folio may probably be attributed to the pleasant appearance of the manuscript as it was originally delivered to the editor or the printer. That it was selected to be printed first in the Folio after it was prepared is more natural than that it was commissioned to be transcribed in a special way so it could serve as a model. Jaggard [the printer] would not have started work before most of the copy, at least for the comedies section, was in his hands. Therefore, Crane must already have prepared transcripts of the first few plays. Further, if *The Tempest* was intended as a model, it cannot have been a good one; *Two Gentlemen of Verona* and *Merry Wives* have significantly different styles of stage-directions. If *The Tempest* was prepared after *Two Gentlemen of Verona*, *Merry Wives* and *Measure for Measure*, for which some evidence may be found, there would have been no point in requiring the scribe to make a special task of the transcript, and he did not accept it as a model for the preparation of his last Folio transcript, *Winter's Tale*. In fact, there is no evidence that Crane prepared copy for *The Tempest* in any special way, and the peculiarities of the other comedies which contrast with *The Tempest* may be attributed to the influence of his copy on him and to his own variability. However, Pollard's view that the publishers chose a clean manuscript of an unpublished play to introduce the Folio is quite plausible.[2]

If the last speculation is admitted, it means that the decision to place *The Tempest* first was not even necessarily made by Heminge and Condell. All this reduces the evidentiary value of the play's place in the Folio to practically nothing.

[1] Ibid., p. 418.
[2] *Crane*, pp. 107–8. I have expanded Howard-Hill's abbreviations of the play titles.

The relation between Crane's text and the work of the compositors has received a good deal of study, though we inevitably remain fairly unenlightened about the details of Crane's copy.[1] It is clear that compositors varied in the degree to which they followed the accidentals and the arrangement of text in their copy, and, in the absence of a surviving manuscript, the question of how much of *The Tempest*'s textual presentation can be attributed to Crane, to the compositors, and to Shakespeare's original can only be guessed at. The heavy punctuation and certain uses of the apostrophe and hyphen are characteristic of those Crane manuscripts that do survive, and several idiosyncratic spellings (e.g. 'Princesse' for 'princes', 1.2.173; 'firrs' for 'furze', 1.1.66) are unquestionably his.[2] The occasional confusion of verse and prose in the text is a more significant matter, not only in those speeches of Caliban's that are set as prose but have the rhythms of blank verse (e.g. at 2.2.154–8 and 161–6), but more puzzlingly when bits of Stephano's and Trinculo's dialogue appear as verse (e.g. 2.2.139–41, 3.2.47–8), or in the anarchic prose of Act 1, Scene 1 where an expostulation of the Boatswain's (13–14) is printed as verse, as is a passage described in the stage direction as '*a confused noise within*' (60–2), but which—on the page at least —is metrically perfectly regular. Some mislineation can be ascribed to the compositors, who in order to fit the text into the space allowed for it when the copy was originally cast off were occasionally required to expand or contract passages as the type was set. However, Howard-Hill writes: 'Crane's habitual use of minuscules at the beginning of lines, mingled with emphasis capitals, might well have obscured the distinction of prose and verse at some points.'[3] Roberts adds that Crane 'often wrote prose in segments which did not fill out lines and therefore looked like verse',[4] and extant Crane texts do not extend prose further to the

---

[1] Howard-Hill's preliminary investigations are recorded and tabulated in *Crane*.

[2] Roberts writes: 'Crane's manuscripts show conclusively that for him "Princesse" was a variant spelling of the male plural and the singular possessive "Princes". A striking use of this alternate spelling occurs in his manuscript of *Sir John Van Olden Barnavelt*, where he writes with clear male reference: "the men you make so meane, so slight account of . . . | are Princes, powerfull Princesse, mightie Princes. . . ."' (p. 228). The *OED* lists 'firrs' as one of a number of possible spellings in the period; it is the form always used by Crane.

[3] *Crane*, pp. 109–10.

[4] Roberts, p. 225.

right of the page than verse, as is customary in other theatrical manuscripts.

What all this means is that there are a number of variables between what Shakespeare wrote and the text that appears in the Folio: the compositors and Ralph Crane, and perhaps also, if Roberts's second hypothesis is correct, the fair copy of Shakespeare's original from which Crane may have been working. Howard-Hill's warning is salutary: 'resort to theories of revision, abridgement, and marginal insertions should not be made until these influences have been allowed for.'[1]

There seems to me only one place in the play where a strong argument for revision can be made, at 1.2.301–4. It is a passage that every editor since the time of the Second Folio has felt a need to emend, and its problems appear to me most easily explicable by assuming a brief cut in the original text, though whether the cut was a feature of Shakespeare's foul papers or of some subsequent revision must, of course, remain finally indeterminable. Such arguments are inevitably subjective, and I offer this one only as a working hypothesis.

The play has in this century been subjected to two theories of wholesale revision. Dover Wilson found the tightness of its construction and its observation of the unities suspect, and postulated an earlier version in which the action covered twelve years, with the first acts of the play dramatizing the events recounted in Prospero's monologue to Miranda in Act 1, Scene 2. This structure is clearly modelled on *The Winter's Tale*, but there is no reason to believe that Shakespeare ever considered it an appropriate one for *The Tempest*. Dover Wilson's thesis was rebutted in detail by Chambers,[2] and has never gained any serious support. There have also been a number of attempts, most notably in this century by Dover Wilson and H. D. Gray,[3] to show that the masque is an interpolation, possibly non-Shakespearian, designed to make the play appropriate to the celebration of the wedding of James I's daughter Elizabeth to the Elector Palatine in February 1613: *The Tempest* was performed a second time at court, along with thirteen other

---

[1] *Crane*, p. 110.

[2] 'The Integrity of *The Tempest*', in *Shakespearean Gleanings* (Oxford, 1944), pp. 76–97.

[3] 'Some Indications that *The Tempest* was Revised', *Studies in Philology*, 18 (1921), pp. 129 ff.

plays, during the season preceding the marriage.[1] The argument requires us to assume first that the play was a much more significant part of the celebration of that wedding than in fact it was, and second that the masque is structurally or stylistically inappropriate to the play as a whole. The masque certainly is stylistically different from the dramatic dialogue of the play, but, since it is a masque, there is no reason why it should not be; and it is thematically and structurally an essential part of the drama. It is difficult to see the argument for interpolation as anything but circular: the only serious evidence adduced for it is the fact that the play was once performed before royal fiancés. The stylization of the masque's action and the relatively stiff formality of its verse are, moreover, quite in keeping with other Shakespearian allusions to the masque, such as Hymen's ceremony in *As You Like It*, Act 5, Scene 4, or Posthumus' vision in *Cymbeline*, Act 5, Scene 4— another passage that has often been claimed to be uncharacteristic of Shakespeare, and therefore spurious. But the point is that it is written in a style appropriate to a dream-vision, as the language of Prospero's entertainment is appropriate to a masque. The arguments are again effectively refuted by Chambers.[2] Chambers's own suggestion that the masque may be celebrating another wedding in 1611 of which we have no record is sensibly dismissed by Kermode, who gives a good summary of the debate and of the issues involved.[3]

The Date. The earliest reference to *The Tempest* is a record in the Revels Accounts of a performance at court on 1 November 1611:

> Hallowmas nyght was presented att Whitehall before
> ye kinges Maiestie a play Called the Tempest.[4]

There is no reason to assume that this was the first performance, but there is good evidence that it was at least an early one. The play is almost certainly indebted to a letter of William Strachey describing the voyage of Sir William Somers to Virginia in the summer of 1609, during which his ship was driven off course and shipwrecked in Bermuda. Strachey's letter is dated from Virginia 15

---

[1] Chambers, *William Shakespeare*, ii. 343.
[2] See p. 44, n. 1. Irwin Smith doggedly pursues the question in 'Ariel as Ceres', *Shakespeare Quarterly*, ix, 3 (1958), pp. 430–2.
[3] Introduction, pp. xv–xxiv.
[4] Chambers, *William Shakespeare*, ii. 342.

July 1610, and would have reached England at the earliest in the beginning of September.[1] Two pamphlets published later in the same year dealt with the Virginia company and with Somers's expedition: *A True Declaration of the Estate of the Colony in Virginia*, the Company's own report of the state of its affairs, and Sylvester Jourdan's *A Discovery of the Bermudas*—Jourdan had accompanied Somers on his voyage. These accounts may also have been in Shakespeare's mind;[2] at any rate, the recent events they describe certainly were. All this suggests that the earliest Shakespeare could have been working on *The Tempest* was in the last months of 1610.

Discussion of the play's date has been complicated by the assumption persisting since the early nineteenth century that *The Tempest* was Shakespeare's last play.[3] A corollary thesis, that *The Tempest* was designed as Shakespeare's farewell to the stage, has strongly conditioned critical attitudes to it. For the moment, we may deal simply with the facts: the first of these assumptions is demonstrably wrong. We know that *Henry VIII* was a new play when it was produced at the Globe in 1613, so it is later than *The Tempest* by at least two years. Shakespeare's two collaborations with Fletcher, *The Two Noble Kinsmen* and the lost *Cardenio*, were also written after *The Tempest*. It has been argued on the basis of internal evidence that *Henry VIII* is also a collaboration, and that therefore *The Tempest* is Shakespeare's last *solo* performance. This may or may not be true, but its significance has less to do with the matter of Shakespearian chronology than with the question of whether the playwright intended the play as a valediction to his career; and the argument here is circular.

There is, in fact, not even any way of determining chronological priority between *The Tempest* and *The Winter's Tale*. Simon Forman recorded in his journal that he saw *The Winter's Tale* at the Globe on 15 May 1611, and the play seems to include an allusion to Ben Jonson's masque *Oberon*, performed at court on New Year's Day 1611.[4] Kermode tries to make something of the fact that Forman

---

[1] See above, p. 32.

[2] Chambers believes Shakespeare 'doubtless used' them: *William Shakespeare*, i. 492.

[3] Malone in the first version of his chronology (1778) set the date at 1612, but believed that *Twelfth Night* was later. This remained his opinion in all versions of the chronology through 1813. The next (and final) version of 1821, however, gives the date as 1611, and puts the play last.

[4] See J. H. Pafford's Arden *Winter's Tale* (1963), pp. xxi–xxii.

'apparently did not see *The Tempest*' at the Globe in the spring season of 1611, when, in addition to *The Winter's Tale*, he saw *Cymbeline, Macbeth*, and a non-Shakespearian *Richard II*.[1] But this tells us nothing about the date of *The Tempest*. It may mean that the play had not yet been produced; but it could as easily mean that Forman saw *The Tempest* at the Globe but did not record the fact, or that *The Tempest* was performed at the Globe but Forman did not see it, or that *The Tempest* was performed that season, but at the Blackfriars, not at the Globe. The most we can say is that the evidence supports a date of late 1610 to mid-1611, and that Shakespeare was writing the play just after, or just before, or at the same time as *The Winter's Tale*.

The Play on the Stage

We can say with reasonable confidence that before the closing of the theatres *The Tempest* was performed at the Blackfriars as well as at court. There is no evidence that it was performed at the Globe, but no reason why it could not have been. If surviving allusions are evidence of popularity, it was not among Shakespeare's best-known works, far less often cited before 1650 than *Hamlet, Romeo and Juliet*, the *Henry IV* plays, *Othello, Richard III, Richard II*, and *A Midsummer Night's Dream*, and about on a par with *Julius Caesar, Macbeth, Much Ado About Nothing, The Merchant of Venice, Pericles, Love's Labour's Lost, The Comedy of Errors*, and *The Winter's Tale*.[2]

The play was first revived after the reopening of the theatres on 7 November 1667, in a version for the most part by Davenant, with some additions by Dryden. *The Tempest, or the Enchanted Island* includes less than a third of Shakespeare's text. Prospero's role is radically reduced, and Miranda and Caliban are provided with sisters, Dorinda and Sycorax. Davenant also introduces, as a parallel to the woman who has never seen a man, a youth named Hippolito, who has never seen a woman. The comic scenes are

[1] Ibid., p. xxii. Indeed, *Cymbeline*, too, may be a work of 1611: Forman's undated entry from the spring or summer of that year constitutes our only clear terminus for the play. There is no external evidence to support Malone's date of 1609, which has become canonical.

[2] *The Shakespeare Allusion Book* cites nine allusions before 1649. The thirty-one allusions recorded between 1650 and 1700 (and G. E. Bentley, in *Shakespeare and Jonson*, adds two more) testify to the great popularity of the Davenant/Dryden and Shadwell versions.

greatly elaborated, Antonio's role diminished, and Sebastian eliminated completely. A good deal of music and dance is added, though Prospero's masque is omitted. The production was very popular. Pepys saw the first performance and noted the presence of the King and court in a 'mighty full' house. He thought it 'the most innocent play that ever I saw', judged it to have 'no wit, but yet good, above ordinary plays', and he especially admired an echo song that had been included for Ferdinand and Ariel.[1] Less than a week later he saw it again, and this time found it 'very pleasant, and full of so good variety that I cannot be more pleased almost in a comedy', though he also found 'the seaman's part a little too tedious'.[2] He saw the play five times in four months.

Davenant's *Tempest* has earned little but scorn from historians of the theatre,[3] but judged on its own merits it is not contemptible, and as a commentary on Shakespeare's play it is enlightening in many of the ways that a good parody can be. The insistent parallelisms and repetitions in Davenant's plot are an elaboration— Davenant might have said a realization—of elements that are implicit in the original, and Davenant's Hippolito represents a genuinely interesting idea about the play. His story acts out, and ultimately disarms, Prospero's most profound fear in Shakespeare's text, of sexuality and the inexorable move towards death. Hippolito is the infant Duke of Mantua, bequeathed by his dying father to Prospero's care. Through astrological calculations, Prospero has learned that the boy will die if he sees a woman.

The relationship between burgeoning sexuality and impending death is never explained further than this, and Davenant employs it, of course, as a comic device, not a potentially tragic one. Prospero's astrological vision has prompted him to keep Hippolito imprisoned, safely hidden from the sight of his two daughters, and the youth has thereby been preserved not only from death, but also from life. The role in the first production was taken by a woman,

---

[1] *The Diary of Samuel Pepys*, eds. Robert Latham and William Matthews, 11 vols. (1970–83), viii. 521 2.

[2] Ibid., p. 527 (13 November).

[3] Two notable exceptions are Jocelyn Powell's enthusiastic and scholarly account of the original staging, in *Restoration Theatre Production* (1984), chap. 4, and John Russell Brown's interesting and sympathetic discussion of the Old Vic revival in June 1959 (*Shakespeare Survey 13* (Cambridge, 1970), pp. 137 ff.). For a brilliant discussion of the political implications of Davenant's revision, see Katharine Eisaman Maus, 'Arcadia Lost: Politics and Revision in the Restoration *Tempest*', *Renaissance Drama*, NS. 13 (1982), 189–209.

initially (as Dryden's Prologue apologetically explains) merely for lack of a suitable young man; but the resulting performance was so successful that Hippolito thereafter remained a breeches part, adding to the play a titillating overtone of sexual ambiguity.[1]

For all its obvious burlesque qualities, Hippolito's sub-plot embodies an astute and accurate critical perception. As an elaboration of Shakespeare's text, it is also unquestionably literalistic and reductive; but its success in its own terms is a historical fact: the inevitable romance of Hippolito and Miranda's sister Dorinda remained a part of *The Tempest* until well into the nineteenth century.

The Restoration *Tempest* had its greatest and most continuous success after 1674, when Shadwell revised Davenant's and Dryden's text into an opera. Some of the vocal music, by John Bannister and Pelham Humphrey, was retained, new songs were set by Pietro Reggio and the gentleman amateur James Hart, new dances were composed by the Master of the Queen's Music, G. B. Draghi, and the instrumental music was by Matthew Locke.[2] It was produced with elaborate scenic machinery, and became one of the great theatrical spectacles of the age. John Downes, the bookkeeper and prompter of Davenant's company for over forty years, writing in 1706, recalled 'particularly one Scene Painted with *Myriads* of *Ariel* Spirits; and another flying away, with a Table Furnisht out with Fruits, Sweet meats, and all sorts of Viands', and he adds, 'all was things perform'd in it so Admirably well, that not any succeeding Opera got more Money'.[3] *The Tempest* in this form became a staple of the theatrical repertory. Purcell provided Dorinda with a new song for a revival in 1690, and at Drury Lane the play was produced in six of the first nine seasons of the eighteenth century—an advertisement in 1707 assures the public that the play will be performed with 'all the Original Flyings and Musick'.[4] Between 1710 and 1732, when the theatre was under the

[1] Powell argues, however, that Dryden was being disingenuous, and that the role was always intended for a woman. See *Restoration Theatre Production*, p. 72.

[2] See Roger Fiske, *English Theatre Music in the Eighteenth Century* (Oxford, 1973), pp. 29–31 and 244–5, and Powell, *Restoration Theatre Production*, p. 206, n. 4. The score underwent numerous revisions and additions, and by the mid-eighteenth century was being attributed entirely to Purcell; it was published as his about 1790. The attribution is no longer accepted: see Fiske, p. 29.

[3] *Roscius Anglicanus* (1708), pp. 34–5.

[4] Cited in *The London Stage 1660–1800, Part 2, 1700–1729*, ed. Emmett L. Avery (Carbondale, 1960), p. 140.

management of Cibber, Wilks, Dogget, and Barton Booth, the opera was revived in twenty of the twenty-three seasons. Cibber notes in his diary for 1712 that it brought in, in the first six days of its run, 'the greatest profit in so little time had yet been known in my memory'.[1]

It was not until 1746 that Drury Lane offered (according to the announcement) the play 'As written by Shakespeare, never acted there before'.[2] James Lacy's production, however, retained a fifth-act masque of Neptune and Amphitrite from Shadwell. This played six performances, a run so modest that in the next year, when Garrick took over the management of the theatre, he returned to the operatic version. In 1756 Garrick presented a new opera based on *The Tempest*, composed by John Christopher Smith, a pupil of Handel's. Smith abandoned most of Davenant's revisions, but retained only the bare essentials of Shakespeare's plot.[3] At last, in 1757, Garrick presented Shakespeare's play. The alterations consisted mostly of cuts (over four hundred lines: almost half of those omitted were in Act 2, Scene 1), and in this form *The Tempest* remained in the Drury Lane repertory until Garrick's retirement in 1776.

Sheridan, in the following year, the first of his reign at Drury Lane, retained Garrick's text but reintroduced both the masque of Neptune and Amphitrite and the 'Grand Dance of Fantastic Spirits' which inaugurates Shadwell's disappearing banquet scene. The production had new settings by de Loutherberg, and was criticized in *The Westminster Magazine* for endeavouring 'to throw an enchantment suited to the childish taste of the present times' over Shakespeare, and for its emphasis on music and dancing—for looking too much like a Christmas pantomime.[4] Machinery was still a crucial element in the play's popularity, and for the next decade the theatre's advertisements made much of the spectacular storm effects, though, in response to complaints from habitual

---

[1] *An Apology for the Life of Colley Cibber*, ed. B. R. S. Fone (Ann Arbor, Mich., 1968), p. 306.

[2] *The London Stage, Part 3, 1729–1747*, ed. Arthur H. Scouten (Carbondale, 1961), p. 1215.

[3] Garrick disclaimed responsibility for the text, but George Winchester Stone argues that he did in fact write it. There is a summary, and a description of the production, in Stone's 'Shakespeare's *Tempest* at Drury Lane', *Shakespeare Quarterly*, 7 (1956), pp. 2–5.

[4] Cited by Odell, i. 431.

latecomers, the public was informed that 'the Storm Scene will (by desire) begin the 2nd Act'.[1] For four seasons between 1776 and 1779 London theatre-goers could also see a competing production of the old operatic *Tempest* at Covent Garden. This had only six performances in all, and was not successful enough to keep in the repertory.

But the move back to Shakespeare's text was only temporary. In 1789 Drury Lane presented a new *Tempest* by John Philip Kemble, which restored Hippolito, Dorinda, and their Restoration colleagues, and much of the spectacle and music of Davenant, Dryden, and Shadwell. In this form the play continued to be enormously popular, and was revived in almost every year of Kemble's tenure at the theatre.

Kemble moved to Covent Garden in 1803. In 1806 he produced another *Tempest*, and for the first time played Prospero himself. The operatic elements were now drastically reduced, though the plot was still substantially Davenant's. This version of the play remained in repertory until 1817, and became the standard acting text for the next quarter of a century. Hazlitt, reviewing a performance in 1815, was outraged by it. The performers he declared mediocre (the Prospero he saw was not Kemble but Charles Mayne Young), and the text was 'travestie, caricature, any thing you please, but a representation'. He was particularly offended to find the 'anomalous, unmeaning, vulgar and ridiculous' Restoration revisions still firmly in place, 'the common-place, clap-trap sentiments, artificial contrasts of situations and character, and all the heavy tinsel and affected formality which Dryden had borrowed from the French school'. The whole experience almost brought him 'to the resolution of never going to another representation of a play of Shakespear's as long as we lived.'[2]

Hazlitt's was still, however, for some years to come, a minority opinion. It was not until 1838 that Macready announced a production of the play at Covent Garden 'in the genuine text of the poet'.[3] The claim was, by the standards of theatrical advertising,

---

[1] The notice was included for all performances beginning 20 September 1777. See *The London Stage, Part 5, 1776–1800*, ed. Charles Beecher Hogan (Carbondale, 1968), p. 114.

[2] *The Examiner*, 23 July 1815, in *Complete Works*, ed. P. P. Howe, 20 vols. (1930–4), v. 234–7.

[3] Macready had played Prospero to John Emery's Caliban in an operatic *Tempest* at Covent Garden in 1821. These were the only spoken roles; the rest were sung.

reasonably accurate: Hippolito, Dorinda, and the rest have once again disappeared, and the play is recognizably Shakespeare's, though a number of songs are retained from the opera for Ariel, played by Priscilla Horton, who sang them while flying about the stage on wires. The first scene was done in a pantomime that was ridiculed by *John Bull* ('a mimic vessel is outrageously bumped and tossed about on waves that we can liken to nothing save tiny cocks of hay, painted green, and afflicted with a spasm'),[1] and a great deal of scenic machinery was employed throughout. From this time on, though the spectacular aspects of *The Tempest* remained a major attraction, the performing version of the play was, despite cuts and the occasional reappearance of Neptune and Amphitrite, firmly based on Shakespeare.

The stage history of *The Tempest* from the Restoration to the beginning of Victoria's reign exhibits a remarkable coherence. Except for two decades under Garrick at Drury Lane, music and machinery were its essence, and the text was, in varying degrees, that of Davenant, Dryden, and Shadwell. What is especially notable in this history is the separateness of the performing and the editorial traditions, which intersect only rarely and relatively briefly. The eighteenth century saw the most serious and continuous effort before the present to produce an accurate and authentic text of Shakespeare; but that was for the library and the scholar. No producer until Garrick ever thought of that authentic text as the one that should or could be *The Tempest* of the repertory. Even Garrick's text cut about 20 per cent out of the play, and the two decades in which it could be seen in Drury Lane constitute only a brief moment in a very long history. Critics from Rowe to Hazlitt called attention from time to time to the travesty the stage had made of Shakespeare's masterpiece, and popular editions of the plays filled the bookshelves, and helped to form the taste, of every literate family in England. Nevertheless Davenant's *Tempest* held the stage.

The play's divided textual history naturally had a profound effect on developing interpretative traditions in the theatre, which, as one would expect, diverged as significantly from the growing critical and editorial tradition as Davenant's and Dryden's text did from

---

[1] Quoted in Odell, ii. 218.

11 (*left*). Ariel as dragonfly: Ann Field at Drury Lane, 1778.
12 (*right*). Ariel as angel: Julia St. George at Sadler's Wells, 1847.

Shakespeare's. In the case of Ariel, however, even Davenant and Dryden proved unsatisfactory guides to performance practice. Ariel had been a male role throughout the seventeenth century. Indeed, Davenant's Ariel is far more explicitly male than Shakespeare's—he is provided with a consort, a female spirit named Milcha—and Dryden's Prologue goes to some pains to point out that in the new theatre, except in the unavoidable case of Hippolito, men are men and women are women. By the early eighteenth century, however, Ariel, like Hippolito, had become exclusively a woman's role, usually taken by a singer who was also a dancer, and so it remained until the 1930s.[1] In this form, Prospero's servant was the central figure in an increasingly elaborate series of operatic and balletic spectacles (see Figs. 11 and 12).

Caliban's character underwent a more complicated change. Dryden, writing in 1679, could find no redeeming elements in him:

[1] The first male Ariel after the Restoration was the dancer Leslie French, in a production at the Old Vic in 1930. See below, p. 77.

he has all the discontents and malice of a witch, and of a devil, besides a convenient proportion of the deadly sins; gluttony, sloth, and lust, are manifest; the dejectedness of a slave is likewise given him, and the ignorance of one bred up in a desert island. His person is monstrous, [as] he is the product of unnatural lust; and his language is as hobgoblin as his person. . . .[1]

This account of Caliban is deaf to the poetry, but is closer to Shakespeare than to Davenant. Davenant's Caliban is far less malign, a bumptious clod, engagingly protective of his sister Sycorax, and even exhibiting an intermittently affectionate nature. On the stage, the role in the eighteenth century tended to stress the comic and burlesque, rather than the malevolent and threatening, and served as a foil for the more standard professional comedy of Trinculo and Stephano. Leigh Hunt, describing John Emery's Caliban in Kemble's 1806 Covent Garden production, records the beginnings of a return to an earlier, more malign, interpretation, and looks ahead to the potentially tragic figure of the latter half of the century:

he . . . approaches to terrific tragedy, when he describes the various tortures inflicted on him by the magician and the surrounding snakes that 'stare and hiss him into madness.' This idea, which is truly the 'fine frenzy' of the poet and hovers on that verge of fancy beyond which it is a pain even for poetry to venture, is brought before the spectators with all the loathing and violence of desperate wretchedness: the monster hugs and shrinks into himself, grows louder and more shuddering as he proceeds, and when he pictures the torment that almost turns his brain, glares with his eyes and gnashes his teeth with an impatient impotence of revenge.[2]

The nature of Prospero's magic, too, went through significant alterations. Arthur Colby Sprague has succinctly charted their course: the spirits at Prospero's command are described in

---

[1] Preface to *Troilus and Cressida*, in *Essays*, ed. W. P. Ker (Oxford, 1900), i. 219–20.

[2] Leigh Hunt, *Critical Essays on the Performers of the London Theatres* (London, 1807), p. 111. Hunt's review of the production in *The News* is reprinted in the Appendix of the same volume, pp. 30 ff. Hazlitt, who saw Emery in the role in 1815, disagreed. Acknowledging that 'he is indeed, in his way, the most perfect actor on the stage', Hazlitt nevertheless complained that 'he has nothing romantic, grotesque, or imaginary about him', and concluded that 'Mr. Emery had nothing of Caliban but his gaberdine, which did not become him' (*Complete Works*, v. 236; see above, p. 68, n. 2).

Davenant and Shadwell simply as 'fantastic'. By the mid-eighteenth century they have become explicitly diabolical, and Prospero has developed Faustian overtones: in 1757, Horace Walpole refers to 'the devils . . . that whisk away the banquet' in Act 3, Scene 3, and Bell's theatre edition of 1773 directs that 'Two Devils rise out of the stage with a Table decorated'. Kemble's 1806 text was at least alluding to Shakespeare in calling instead for 'Three spirits, in the shape of Harpies', but Charles Kean, in his 1857 production at the Princess's Theatre, replaced all such monstrosities with a group of 'naiads, wood-nymphs, and satyrs . . . bearing fruit and flowers'—*The Athenaeum*'s reviewer made the familiar complaint that this staging was 'too much like a transformation scene in a pantomime'.[1]

As the criticism suggests, the return to Shakespeare's text was accompanied by no diminution of spectacular effects. *The Tempest* was still a machine-play *par excellence*, and *The Saturday Review* observed of the acting in Kean's production that 'there is not much room for it'.[2] A look at Kean's prompt-book reveals that this was literally true: the text was heavily cut,[3] but the performance ran five hours, almost double the fictive time represented in the action.

Such productions prompted theatrical as well as critical reactions. Samuel Phelps's Sadler's Wells production of 1847, though it too was heavily cut and employed such complex scenic effects as burning rocks and a magic fire surrounding Prospero during the banquet scene, had been especially praised for its emphasis on the text (*The Times*'s reviewer called it the 'best combination of Shakespeare and scenery'[4]). Frank Benson at the Shakespeare Memorial Theatre in Stratford in 1891 radically simplified the staging, though he dealt with the problems of the opening

---

[1] *Shakespeare and the Actors* (Cambridge, Mass., 1944), p. 43.

[2] 4 July 1857.

[3] For example, Kean not only made the standard cuts of all the dialogue in the opening scene, most of the exchanges between Alonso, Antonio, Sebastian, and Gonzalo in Act 2, Scene 1, and much of the masque, but also omitted 160 lines from Prospero's dialogue with Miranda and Ariel in Act 1, Scene 2, much of Ferdinand's and Miranda's log-bearing scene in Act 3, Scene 1, and large sections of Ariel's encounter with the three men of sin and of Prospero's renunciation speech and epilogue. *The Tempest*, it should be remembered, is a very short play: only *The Comedy of Errors* is shorter in the Shakespeare canon.

[4] John Oxenford in *The Times*, quoted in Samuel Phelps *et al.*, *The Life and Life-Work of Samuel Phelps* (1886), p. 222.

scene by omitting it,[1] and had Ferdinand enter at the end of a silver thread drawn by two cupids. In 1897 William Poel's Elizabethan Stage Society presented the play on an open stage without scenery, and banished the orchestra in favour of a musical score by Arnold Dolmetsch using only the pipe and tabor. These productions were not received uncritically, but both were highly praised for allowing the words to carry the play.

The most interesting interpretative developments in Victorian productions increasingly focused on the figure of Caliban. As the age progressed, he grew more malign, but also less diabolical, more elementally human, at once more richly comic and more deeply tragic. Trevor R. Griffiths, in an essay tracing the growth of Caliban into an embodiment of republican and anti-slavery sentiments, points out that in productions throughout the first two-thirds of the century, the final scenes of the play had concentrated on Prospero and Ariel, 'often with spectacular scenes of Ariel's flight and Prospero's departure on the restored ship. In such versions, no specific provision was made either for leaving Caliban on the island or taking him away.' But in John Ryder's 1871 production at the Queen's Theatre, as the ship sailed off, the island became 'the sole charge of Caliban, who as the curtain descends lies stretched upon the shore basking in the rays of the setting sun. Clearly this was a Caliban glad to be left behind in charge of his island.'[2] Benson in his Stratford production made Caliban a Darwinian missing link, the embodiment of inchoate but aboriginal and essential humanity[3]—a recapitulation in Victorian terms of Harriot's discovery of the ancient British past in the savages of Virginia.[4] Benson based his movements on the behaviour of great apes, which he had studied at length, and gave an astonishingly athletic performance that audiences found both disturbing and hilarious. Beerbohm Tree's 1904 *Tempest* greatly expanded Caliban's role, emphasizing his sensitivity to music, and including

---

[1] It was restored in his 1900 production, and staged with the utmost economy: 'a lantern swung back and forth on an otherwise dark and bare stage, while the actors tumbled about' (Carey M. Mazer, *Shakespeare Refashioned* (Ann Arbor, Mich. 1981), p. 72).

[2] ' "This Island's Mine": Caliban and Colonialism', *Yearbook of English Studies*, 13 (1983), p. 164.

[3] The interpretation was based on Daniel Wilson's *Caliban: The Missing Link* (Toronto, 1873).

[4] See above, pp. 34–5.

13. Sir Herbert Beerbohm Tree as Caliban dancing, 1904, drawing by C. Harrison, published in *The Sketch*.

elaborate dance and pantomime sequences for him (see Fig. 13). Tree's description of the concluding tableau, with Caliban at its centre, has already been cited.[1]

Tree was particularly alive to the new possibilities that electricity brought to the theatre, and his *Tempest* was technologically very sophisticated. For the most part he abandoned heavy machinery in favour of complex and subtle lighting effects, and he wrote in his souvenir programme that 'of all Shakespeare's works, *The Tempest* is probably the one which most demands the aids of modern stagecraft'. In this belief he was, of course, less the apostle of a new theatre than the latest heir of Davenant and Shadwell, and his was the last of the great spectacular *Tempests*.[2] In contrast, Ben Greet's Old Vic production of 1914 employed no special scenic effects, and had a running time of just two hours. The operatic *Tempest*, however, was still in evidence: the requisite atmosphere of illusion and enchantment was created by an introductory ballet of spirits—the

[1] See above, p. 26.
[2] For a detailed account of the production, see Mary M. Nilan, ' "The Tempest" at the Turn of the Century', *Shakespeare Survey* 25 (Cambridge, 1972), pp. 118 ff.

music was that of the sixteen-year-old Sir Arthur Sullivan, com posed half a century earlier as a student at the Leipzig conservatory. And, perhaps now inspired as much by the operatic precedents of Cherubino and Octavian as by Christmas pantomimes (to say nothing of Davenant's Hippolito), Greet's Ferdinand was played by Sybil Thorndike.

Five years later William Bridges-Adams, in his first season at Stratford, undertook to return the play to history. Praising Poel's efforts, he went a step further by linking his production to the particular occasion for which he believed *The Tempest* had been specifically designed. He employed a gauze curtain which displayed the portraits and arms of the Elector Palatine and Princess Elizabeth, and attempted in his staging to reproduce, in a very simplified manner, the effect of a Jacobean court masque. As a theatrical event, this production was admired primarily for the elegant economy of its presentation, but it is also an important manifestation of a new movement in the critical history of the play : Ernest Law's Shakespeare Association Pamphlet *Shakespeare's 'Tempest' as Originally Produced at Court* was published in the following year, 1920, and in 1921 Dover Wilson's New Cambridge edition of the play appeared, announcing in its Introduction that 'we may take it almost for a certainty that—in whatever previous form or forms presented—this play *as we have it* was the play enacted at Court to grace the Princess Elizabeth's betrothal'.[1]

Bridges-Adams conceived his *Tempest*, in part at least, as the re-creation of a historical moment. The notion that theatrical productions require the validation of history has appeared intermittently since the mid-eighteenth century. In the 1740s and 1750s, dramatic critics began to complain of anachronisms in stage settings and dress, and in the 1760s both Drury Lane and Covent Garden were advertising plays with historically accurate costumes. The fashion came and went, but by the 1820s the researches of J. R. Planché had established a norm of historic clothing for the stage. For the eighteenth century, the movement found its most enduring monument in the engraved paintings of John Boydell's Shakespeare Gallery, whose announced intention was to create, through the depiction of Shakespearian subjects, a native school of history painting.

[1] p. xlvii.

John Philip Kemble had produced a *Hamlet* in Elizabethan dress in 1783. The history in this case was not Hamlet's but Shakespeare's, and Kemble was a pioneer in the effort to re-create in a modern playhouse the experience of Shakespeare's audience. For Victorian and Edwardian producers, however, the history was almost invariably that of the play's subject matter. The text was assumed to be transparent; the play was a representation, and what was authentic in it was what was represented. Consistency and accuracy were the cardinal virtues in the treatment of action, dress, and settings—though not, of course, in the treatment of the script.

Dramatic productions today are rarely conceived in so direct a way as re-creations of the past (the film, as the more realistic medium, generally takes over that function), and the pressure towards consistency of all kinds has significantly diminished. Nevertheless, the alliance between theatre and history remains in its way a strong one: modern Shakespeare productions often take possession of the Renaissance, and in doing so find the meaning of the play in the version the past can be made to provide of our own history. The practice has found its most articulate justification in Jan Kott's *Shakespeare Our Contemporary*. And the text remains endlessly malleable: modern directors feel no less free than Kean and Beerbohm Tree to assign Miranda's attack on Caliban to Prospero, to cut the masque, or to replace Prospero's epilogue with 'Our revels now are ended. . . .' For the theatre, the play has never been fully or adequately represented by its text, and its verisimilitude has often been provided not by its dramatic consistency but by the history that is assumed to be speaking through it.

Dover Wilson's account of *The Tempest*'s archaeology is as much part of this movement as Bridges-Adams's 'Jacobean' production at Stratford. But the peculiar attractiveness of Dover Wilson's thesis for literary scholars had less to do with its historical claims than with their explanatory value, their apparent ability to tie the generation of a Shakespeare text to a specific event and historical moment. The theatre at this period, however, like the other visual arts—and like literary criticism a decade or two later—was moving away from this sort of localization and particularity towards the generalizing power of abstraction on the one hand and of contemporaneity on the other. Even in Bridges-Adams's production the historical argument did not preclude a radically simplified staging, with an abstract setting consisting largely of arrange ments of curtains.[1]

If the idea of *The Tempest* as a specifically Jacobean play had little immediate effect on subsequent productions, it must, nevertheless, have contributed to one very significant theatrical innovation: at the Old Vic in 1930, Ariel became a male role again for the first time in over two centuries. There were, no doubt, powerful contemporary models to be found in the London triumphs of Nijinsky and his successors at the Russian ballet; but it would have been an awareness of the Jacobean *Tempest* that enabled directors to find such models relevant to Ariel. Harcourt Williams cast the dancer Leslie French in the part, to John Gielgud's Prospero and Ralph Richardson's Caliban. French was boyish, light, and active, costumed in only a loincloth and a winged hat suggestive of Mercury. Critics praised his energy and the longing for freedom it implied. He played the part again in Robert Atkins's Regent's Park production in 1934.

Ariel was not thereby transformed suddenly and permanently into a male role; but to treat it as such was once again possible in the theatre. Indeed, within a decade the question of the spirit's gender was no more significant than a director wanted to make it for the purposes of a particular interpretation. French was followed in 1934, both at the Old Vic and at Sadler's Wells, by Elsa Lanchester, playing a very art-deco and decidedly feminine Ariel in a silver tunic, wings, and a cape, and wearing lipstick and mascara (Fig. 14). James Agate praised her extravagantly for the lightness and radiance she brought to the part, compared her (perversely, surely) to Nijinsky's Faun, and claimed that 'until Miss Elsa Lanchester the part of Ariel has never been acted'.[2] It is not clear what Agate found so innovative about the performance; but in fact, far from providing a new norm, it may be taken to mark the end of a long interpretive history.[3]

---

[1] Nor did historicism involve any attention to the original text: Bridges-Adams wrote to Robert Speaight in 1953: 'I never knew the Folio gave "Abhorred slave" to Miranda.' (*A Bridges-Adams Letter Book*, ed. Robert Speaight (London, 1971), p. 56.)
[2] Review of 8 January 1934, reprinted in Agate's collection *Brief Chronicles* (1943), p. 21.
[3] Charles Laughton played Prospero, and the production was by Tyrone Guthrie, who called it 'probably the worst production of *The Tempest* ever achieved'. Tyrone Guthrie, *A Life in the Theatre* (New York, 1959), p. 127.

14. Elsa Lanchester as Ariel and Charles Laughton as Prospero
at the Old Vic, 1934.

In 1940 a second male Ariel appeared at the Old Vic, in a
radically new interpretation of the part—arguably the first new
interpretation since the sprite was originally cast as a female dancer two centuries before. Marius Goring (Fig. 15) was

silvered, leaden-coloured almost, but with a quick quivering poise of the
head that belonged neither to bird nor man, but to some weird immortal
spirit imprisoned in mortal toils and painfully eager to be free. In all his
contacts with humans one sensed the unfeeling curiosity of the sprite: not
cruel, but cool and remote, as if the world of humans was as strange to him
as his world to the mortals.[1]

The old Ariel had been a principle of grace and freedom that Prospero must relinquish in returning to the world of responsibility and
mortality; Goring's was instead an alien figure in bondage, eerie
and distinctly non-human. The interpretation was to have a
powerful effect on post-war productions: Ariel was thereafter almost always cast as a man, and a good deal of sentimentality—and
humanity—disappeared from the role.

---

[1] Audrey Williamson, *Old Vic Drama* (Rockcliff, 1948), p. 142. The production
was by Goring and George Devine.

15. Marius Goring as Ariel, Old Vic, 1940.

The major interpretative issue in productions over the past fifty years has centred on the character of Prospero. He had traditionally been presented as a benign wizard, elderly, serene and majestic. His control over himself and over the action, like the ethics of his paternalism, had generally gone unquestioned—James Agate detected a strong element of Father Christmas in Charles Laughton's performance in 1934.[1] The figure in the text, however, is far more ambiguous, and directors started to become interested in the possibilities of a less sentimental Prospero at the same time as they discovered a less human Ariel.

There is, to begin with, the question of Prospero's age. Agate was drawing on the experience of many productions over many years when he called him an 'old codger',[2] but the question is a more open one than it has generally appeared. Prospero refers to himself as old only once, in explaining his interruption of the masque to Ferdinand: 'Bear with my weakness, my old brain is troubled' (4.1.159). This and his brief contemplation of impending mortality, 'Every third thought shall be my grave' (5.1.311), constitute his only allusions to advancing age. How far we want to

[1] *Brief Chronicles*, p. 19; see above, p. 77, n. 2.
[2] Ibid., p. 17.

79

treat these as literal statements will depend on our sense of the play as a whole—the question is similar to the one about Caliban's paternity: are Prospero's assertions to be taken as facts about the world of the play, or only as facts about Prospero? Physiologically he need not be much more than thirty; he has a fifteen-year-old daughter. The past action he recounts, the retirement to his library, the usurpation, obviously would have required more time than a teenage duke would allow for, and the younger brother, too, would have had to be old enough to rule and to effect the usurpation; but how much time one wants to allot to the play's history will depend, again, on one's view of the play as a whole. Prospero's sense of his age has to do with his sense of his power, or of his potency, and these are his magic and his control over other people: Miranda, Ariel, Caliban, and, far less effectively, the ship-wreck victims. Wizards, of course, are conventionally represented as old, but Prospero only *declares* himself old when his daughter is ready to marry. There is surely more dramatic psychology in this than physiology.

For the nineteenth and early twentieth centuries, the impulse to play Prospero as old would have been strengthened by the critical assumption that Prospero is Shakespeare and that *The Tempest* signalled his retirement from the stage. In this regard, it is worth remembering that in 1611 Shakespeare was forty-seven. He may, of course, have thought of himself as old at that age; but Ben Jonson, musing on his stroke at the age of fifty-four, complained that 'thy nerves be shrunk and blood be cold | Ere years have made thee old'.[1] Several recent productions have presented Prospero as vigorously middle-aged, or even youthful. In both Derek Jacobi's performance for the RSC in 1983 and Michael Hordern's in the 1979 BBC version, the sudden claim of age and infirmity came as a powerful surprise.

Gielgud's Prospero, developed through a series of performances from 1930 to 1973, has proved in important ways a normative one for the modern theatre, though it is a norm that many directors have wished to displace. Expressive, intellectual, fastidious, his reading has always been built around the setpieces, and has given great emphasis to the richness and beauty of the verse throughout. At the Old Vic in 1930, at the age of twenty-six, he played the part

[1] 'Ode to Himself' ('Come, leave the loathèd stage . . .'), ll. 45–6.

16 *(left)*. Michael Redgrave as Prospero and Alan Badel as Ariel in Michael Benthall's elegant and fantastic production at Stratford, 1951.

17 *(right)*. Hugh Griffith's Caliban, Stratford, 1951.

18. Prospero (Michael Redgrave) draws the magic circle, Stratford, 1951.

19. Sir John Gielgud as Prospero and Alec Clunes as Caliban in Peter Brook's production at Stratford, 1957.

beardless and modelled his appearance on Dante.[1] In 1940 he again looked no more than his own age, wore a small goatee, and at times used spectacles. He brought to the role 'a certain wry humour and scholastic irony'.[2] The magic of the play seemed his natural element; and in this interpretation, the exercise of Prospero's power expressed itself as a continual retreat into a world of fantasy.

Such a view of the play will always find adherents (see Figs. 16–18). But post-war directors have been on the whole less interested in the escapist aspects of *The Tempest*, and have often allowed the text's ambivalence and inconclusiveness to set the tone. Reviewers generally grumble at this, clearly longing not for a more accurate reading, but for a lost innocence. Michael Hordern at the Old Vic in 1954 frankly emphasized Prospero's bitterness;[3]

[1] John Gielgud, *Early Stages*, rev. edn. (New York, 1976; London, 1979), p. 113.
[2] Williamson, *Old Vic Drama*, p. 140, see above, p. 78, n. 1.
[3] Robert Helpmann directed.

though generally admired, he was also found insufficiently poetic by several critics. Peter Brook, at Stratford in 1957, cast Gielgud as an introspective and obsessive Prospero, isolated and brooding on his wrongs. Brook made the play a projection of Prospero's inner world, with cavernous settings and overgrown, tangled vegetation —a dream world, but, as *The Times*'s critic put it, 'the kind of dream King Lear might have had' (Fig. 19).[1] Most critics found Brook's emphasis on the dark side of the play relentlessly monochromatic; for Roy Walker, though Gielgud was a superb embodiment of Brook's interpretation, 'never have the Ariel songs sounded more dreary, the masque been more eerie, or the scenery more obtrusively scenic'.[2]

It is true that the most familiar attractions of the play, the theatrical magic that is so central an element in its stage history, will not be found in such a reading. Nevertheless, the move away from a benign and serene Prospero has opened up to the theatre a range of possibilities that are fully implicit in the text, or perfectly valid extensions of it (see Figs. 20, 21). Recent productions by Jonathan Miller and Peter Hall provide good illustrations. Miller, in a 1970 production at the Mermaid, based his view of the relation of Prospero to Caliban and Ariel on Octave Mannoni's metaphorical use of these figures in his analysis of the revolt in Madagascar in 1947, *La Psychologie de la colonisation*.[3] Some critics objected that an anachronistic political argument was being imposed on the play, but this is only true in a very limited sense. Such an interpretation need not imply that there are no differences between a Jacobean view of exploration narratives and modern views of the colonialist experience; but it does acknowledge that audiences have changed too, and that the theatre's texts are affected by history. Miller explained his intentions in an interview with Ralph Berry. Mannoni, in Miller's account,

saw Caliban and Ariel as different forms of black response to white paternalism. In Caliban he saw the demoralised, detribalised, dispossessed shuffling field hand and in Ariel a rather deft accomplished black who actually absorbs all the techniques and skills of the white master. . . . I

[1] 6 December 1957, when the production had moved from Stratford to Drury Lane.

[2] *Shakespeare Survey* 11 (Cambridge, 1958), p. 135.

[3] Paris, 1950; translated into English by P. Powesland as *Prospero and Caliban*, (London and New York, 1956).

20. Ben Kingsley as a macho Ariel to Ian Richardson's querulous Prospero in John Barton's production at Stratford, 1970.

21. Trinculo (Norman Rodway), Stephano (Patrick Stewart) and Caliban (Barry Stanton), Stratford, 1970.

wasn't using *The Tempest* as a political cartoon to illustrate the Nigerian dilemma nor as it were to castigate modern colonialism . . . but to use . . . the whole colonial theme as knowledge which the audience brought to bear on Shakespeare's play. They could scarcely avoid thinking of that situation when the two characters were represented as blacks. . . . By doing it in this way I hoped to bring them into a closer relationship with the whole notion of subordination and mastery which I think is one of the things which Shakespeare is talking about with great eloquence in that play. And I think he is also talking about . . . infantilism and about the way in which maturity is only arrived at by surrendering one's claim to control the whole of nature. A child after all arrives at maturity by appreciating the reality principle, . . . simply the understanding that there are certain things over which one has control, and there are most things over which one has no control.[1]

The historical analogy obviously was no impediment here to a very clear perception of the complexity of Prospero's character and of his relation to his two servants.

Peter Hall returned to Jacobean history for a production at the Old Vic in 1973; his *Diaries* include an account of its inception and progress. Hall based his conception on a study of the court masque and the stage designs of Inigo Jones, and he correctly took masques to be not simply theatrical spectacles, but forms and expressions of royal power. To see the play in the context of the masque, then, was also to abandon the traditional Prospero, the serene and detached enchanter. Hall initially approached Olivier to play the part, rather than Gielgud, who, he felt, was too much identified with the standard interpretation: 'I pointed out that Prospero was acted traditionally by a remote old man—an aesthetic schoolmaster who was thinking of higher things, whereas Prospero should really be a man of power, of intelligence, as shrewd and cunning and egocentric as Churchill.'[2] Olivier was interested, and was equally dissatisfied with the limitations of the usual stage Prospero. He proposed an interpretation of his own that would bring the character emphatically down to earth: 'He said he wanted to play the part for comedy, and that Prospero should lecture his daughter in the first scene while shaving. He said he couldn't wear all those whiskers and wigs Prospero always wore. . . .'[3] Olivier, however, proved unable to accept the role. Hall turned with some misgivings

[1] *On Directing Shakespeare* (1977), pp. 34 ff.
[2] *Peter Hall's Diaries*, ed. John Goodwin (1983), p. 12.
[3] Ibid., p. 43.

to Gielgud, worrying that 'he is perhaps too gentle and nice', and determining 'to push him to a harsher area of reality'. Hall wanted Gielgud to do Prospero 'as a William Blake-like figure going through the purgatory of the play. I want the play to happen *to* Prospero, rather than have a sweet, well-spoken old retired vicar meditating on it from the outside'.[1] As rehearsals proceeded, Hall's conception of the role moved increasingly towards the darkest side of Prospero:

Prospero is a man who is contained and careful. He does not reveal himself to the audience. He can barely reveal himself to himself. He is controlled. His passion for revenge is not emotional but puritanical. [Gielgud] shows the agony that Prospero is going through from the very beginning of the play. He should wait until the end. Macbeth has to be played by an actor who is content to act dangerously little for the first half of the play. The technique is the same with Prospero. . . .[2]

Behind this, obviously, is a bold, perceptive, and original reading of the play: to see Prospero in the context of *Macbeth* is to turn the stage history of *The Tempest* on its head.

Such productions seem perverse only in relation to the play's history in the theatre: the text itself provides ample support for them. And indeed, it may be said that the focus of modern productions is increasingly textual in the sense that the play is conceived less as representation (e.g. of a true or consistent history) than as the *presentation* of a text—not as the mirror of a previous action, but as the action itself. Peter Brook, at the Roundhouse in 1968, carried this notion to its logical extreme in a fascinating version of the play, which may serve as our final example. Brook's production was an Artaudesque fantasia designed to explore and release the play's ambivalent, disruptive, and frankly violent energies. What resulted, in the words of Margaret Croyden,

was not a literal interpretation of Shakespeare's play, but a working out of abstractions, essences, and contradictions embedded in the text. The plot was shattered, condensed, deverbalized; time was discontinuous, shifting. . . . Whenever Shakespeare's words were spoken, they were intoned and chanted. Brook tried to strip the play of preconceived language patterns connected with classical interpretations of Shakespeare.[3]

---

[1] Ibid., pp. 43–4.
[2] Ibid., p. 76.
[3] Margaret Croyden, *Lunatics, Lovers and Poets: The Contemporary Experimental Theater* (New York, 1974), p. 246.

Not surprisingly, the energy thereby released was primarily sexual and rebellious, the energy most feared and most powerfully suppressed in the text. Ferdinand and Miranda meet and make love, parodied by Ariel and Caliban, and subsequently by other members of the cast. Prospero undertakes to train and subjugate Caliban by teaching him to speak: the vocabulary consists of 'I', 'you', 'food', 'love', 'master', 'slave'. But the last two words release both Caliban's rebelliousness and his libido; he escapes from Prospero, accomplishes the rape of Miranda, and takes over the island. An orgiastic scene follows, with Prospero as the central victim.

The reconciliation is provided, as it is in the play, by Ariel: he brings ribbons, costumes, bright clothing —material things—to bribe the dog pack. The group breaks into game improvisations, and the scene dissolves into Miranda's and Ferdinand's marriage ceremony. . . . The wedding over, Prospero says, 'I forgot the plot.' The double entendre refers to the actual play and to Caliban's plot. . . . Each actor stops where he is, thinks a moment, and then someone begins the lines from *The Tempest* epilogue: 'And my ending is despair'. . . . The verse, spoken in various rhythms, inflections, intonations, and phrasings, mixes the sounds until everyone fades out, leaving the audience in stillness.[1]

Brook's reading of *The Tempest* here is, obviously, a partial one; though it is arguably no more partial than the stage's traditional sentimental reading. And, once again, it is only in the context of stage history that a reading such as Brook's will be found perverse: it is designed to bring into the theatre a recognition of how powerfully subversive much of the play's energy is, how incompletely it controls its ambivalences and resolves its conflicts.

[1] Ibid., pp. 249–50.

EDITORIAL PROCEDURES

I HAVE followed the procedures set up by the General Editor. These are described in detail in *Henry V*, edited by Gary Taylor (Oxford, 1982), pp. 75 ff. Modernization has followed the principles established in Stanley Wells's 'Modernizing Shakespeare's Spelling', in Wells and Taylor, *Modernizing Shakespeare's Spelling, with Three Studies in the Text of 'Henry V'* (Oxford, 1979). In the Collations, insignificant variations in speech headings (e.g. Anthonio, Ariell) have not been recorded, nor have obvious misprints (e.g. pectlesse for peerless, 3.1.47; starngely for strangely, 5.1.313). Punctuation has been collated only where a significant syntactical question is involved. Since all asides and speech directions (e.g. *to Trinculo*) are editorial, only those that are original to this edition have been collated. All changes to stage directions are noted in the collations, but where the specified action is clearly implied by the dialogue, the change is neither bracketed in the text nor attributed to a particular editor. Disputable alterations are printed within broken brackets (⌈ ⌉).

Shakespeare seems to have relied on at least two of his sources, Montaigne's essay 'Of the Cannibals' and the Strachey letter, not only for verbal details but more broadly for attitudes and models of behaviour. I have, therefore, included in appendices the complete Montaigne essay and more extensive selections from Strachey than is usual. For the same reason, I have reprinted the whole scene of Medea's invocation from Golding's Ovid, rather than merely the excerpt that is directly echoed in the play. The texts are based on the original editions cited, and have been modernized.

Abbreviations and References

The following abbreviations are used in the explanatory notes and collations. The place of publication is London unless otherwise specified.

EDITIONS OF SHAKESPEARE

F	The First Folio, 1623
F2	The Second Folio, 1632
F3	The Third Folio, 1663
F4	The Fourth Folio, 1685
Alexander	Peter Alexander, *Works*, The Tudor Shakespeare (1951)
Barton	Anne (Righter) Barton, *The Tempest*, The New Penguin Shakespeare (Harmondsworth, 1968)
Cambridge	W. G. Clark and W. A. Wright, *Works*, The Cambridge Shakespeare, 9 vols. (Cambridge, 1863–6)
Capell	Edward Capell, *Comedies, Histories and Tragedies*, 10 vols. (1767–8)
Collier	John Payne Collier, *Works*, 8 vols. (1842–4)
Dyce	Alexander Dyce, *Works*, 6 vols. (1857)
Dyce 2	Alexander Dyce, *Works*, 2nd edn., 9 vols. (1864–67)
Grant White	*Works*, 12 vols. (Boston, 1857–66)
Hanmer	Thomas Hanmer, *Works*, 6 vols. (Oxford, 1743–4)
Johnson	Samuel Johnson, *Plays*, 8 vols. (1765)
Kermode	Frank Kermode, *The Tempest*, The Arden Shakespeare, sixth edn. revised, with corrections (1961)
Kittredge	G. L. Kittredge, *The Tempest* (Boston, 1939)
Knight	Charles Knight, *Works*, Pictorial Edition, 8 vols. (1838–43)
Malone	Edmond Malone, *Plays and Poems*, 10 vols. (1790)
Neilson – Hill	William Allan Neilson and Charles Jarvis Hill, *Plays and Poems*, The New Cambridge Edition (Cambridge, Mass., 1942)
Pope	Alexander Pope, *Works*, 6 vols. (1723–5)
Rann	Joseph Rann, *Dramatic Works*, 6 vols. (1786–94)
Riverside	G. B. Evans (textual editor), *The Riverside Shakespeare* (Boston, 1974)
Rowe	Nicholas Rowe, *Works*, 6 vols. (1709)
Rowe 1714	Nicholas Rowe, *Works*, 8 vols. (1714)
Sisson	C. J. Sisson, *Complete Works* (1954)
Steevens	Samuel Johnson and George Steevens, *Plays*, 10 vols. (1773)

Theobald	Lewis Theobald, *Works*, 7 vols. (1733)
Variorum	Horace Howard Furness, *The Tempest*, A New Variorum Edition (Philadelphia, 1892)
Warburton	William Warburton, *Works*, 8 vols. (1747)
Wilson	John Dover Wilson, *The Tempest*, The New Shakespeare (Cambridge, 1921)
Wright	W. A. Wright, *The Tempest*, The Clarendon Shakespeare (Oxford, 1874)

OTHER WORKS

Abbott	E. A. Abbott, *A Shakespearian Grammar*, 3rd rev. edn. (1870)
Allen, *Notes*	[George Allen, I. Fish and II. H. Furness, eds.,] *Notes of Studies on 'The Tempest'*, Minutes of the Shakspere Society of Philadelphia for 1864–65 (Philadelphia, 1866)
Baldwin	T. W. Baldwin, *William Shakspere's 'Small Latine & Lesse Greeke'* (Urbana, Ill., 1944)
Bulloch	John Bulloch, *Studies of the Text of Shakespeare* (1878)
Bullough	Geoffrey Bullough, *Narrative and Dramatic Sources of Shakespeare*, vol. 8 (1975)
Cercignani	Fausto Cercignani, *Shakespeare's Works and Elizabethan Pronunciation* (Oxford, 1981)
Chambers	E. K. Chambers, *The Elizabethan Stage*, 4 vols. (Oxford, 1923)
Chambers, *William Shakespeare*	E. K. Chambers, *William Shakespeare: A Study of Facts and Problems*, 2 vols. (Oxford, 1936)
Cotgrave	Randle Cotgrave, *A Dictionarie of the French and English Tongues*, 2nd edn. (1632)
Crane	T. H. Howard-Hill, *Ralph Crane and Some Shakespeare First Folio Comedies* (Charlottesville, 1972)
Davenant–Dryden	Sir William Davenant and John Dryden, *The Tempest, or The Enchanted Island* (1670), in *Works of John Dryden*, vol. 10 (*Plays*), eds. Maximillian E. Novak and George Robert Guffey (Berkeley and Los Angeles, 1970)
Dent	R. W. Dent, *Shakespeare's Proverbial Language: An Index* (Berkeley and Los Angeles, 1981)
Douce	Francis Douce, *Illustrations of Shakespeare*, 2 vols. (1807)
ES	*English Studies*

Elze	Karl Elze, *Notes on Elizabethan Dramatists, with conjectural emendations of the text*, second series (Halle, 1884)
Greg	W. W. Greg, *The Shakespeare First Folio* (Oxford, 1955)
Golding	W. H. D. Rouse, *Shakespeare's Ovid, Being Arthur Golding's Translation of the Metamorphoses* (1904, repr. 1961)
Grigson	Geoffrey Grigson, *The Englishman's Flora* (1955)
Kökeritz	Helge Kökeritz, *Shakespeare's Pronunciation* (New Haven, 1953)
MLN	*Modern Language Notes*
McIlwain	C. H. McIlwain, ed., *The Political Works of James I* (Cambridge, Mass., 1918)
N.&Q.	*Notes and Queries*
Noble	Richmond Noble, *Shakespeare's Biblical Knowledge and Use of the Book of Common Prayer* (1935)
OED	*The Oxford English Dictionary, being a corrected re-issue . . . of A New English Dictionary on Historical Principles*, 13 vols. (Oxford, 1933), and Supplements 1–2 (1972, 1976)
Odell	George C. D. Odell, *Shakespeare from Betterton to Irving*, 2 vols. (New York, 1920)
Onions	C. T. Onions, *A Shakespeare Glossary*, 2nd edn. (1919), repr. with addenda (Oxford, 1958)
PMLA	*Publications of the Modern Language Association of America*
RES	*Review of English Studies*
Representing Shakespeare	Murray Schwartz and Coppelia Kahn, eds., *Representing Shakespeare* (Baltimore, 1980)
Roberts	Jeanne Addison Roberts, 'Ralph Crane and the Text of *The Tempest*', *Shakespeare Studies 13* (Cambridge, 1980), pp. 213–33
ShQ	*Shakespeare Quarterly*
Sisson, New Readings	C. J. Sisson, *New Readings in Shakespeare*, 2 vols. (Cambridge, 1956)
Tilley	M. P. Tilley, *A Dictionary of the Proverbs in England in the Sixteenth and Seventeenth Centuries* (Ann Arbor, 1950)
Walker	W. S. Walker, *Shakespeare's Versification and its apparent irregularities explained* (1854)

The Tempest

THE PERSONS OF THE PLAY

ALONSO, King of Naples

SEBASTIAN, his brother

PROSPERO, the right Duke of Milan

ANTONIO, his brother, the usurping Duke of Milan

FERDINAND, son to the King of Naples

GONZALO, an honest old councillor

ADRIAN and FRANCISCO, lords

CALIBAN, a savage and deformed slave

TRINCULO, a jester

STEPHANO, a drunken butler

MASTER OF A SHIP

BOATSWAIN

MARINERS

MIRANDA, daughter to Prospero

ARIEL, an airy spirit

IRIS

CERES

JUNO $\Big\}$ personated by spirits

NYMPHS

REAPERS

The scene: an uninhabited island

The Tempest

1.1 *A tempestuous noise of thunder and lightning heard.*
Enter a Ship-master and a Boatswain

MASTER Boatswain!

BOATSWAIN Here, master. What cheer?

MASTER Good—speak to th' mariners. Fall to't yarely, or we
run ourselves aground. Bestir, bestir!   *Exit*
> *Enter Mariners*

BOATSWAIN Hey, my hearts! Cheerly, cheerly, my hearts!
Yare, yare! Take in the topsail. Tend to th' master's
whistle. (*To the storm*)—Blow till thou burst thy wind, if
room enough!
> *Enter Alonso, Sebastian, Antonio, Ferdinand, Gonzalo,*
> *and others*

ALONSO Good boatswain, have care. Where's the master?

Persons of the Play] Names of the Actors, *at end in* F   1.1.7 *To . . . storm*]
This edition; *not in* F   8.1 *Ferdinand*] ROWE; *Ferdinando* F

1.1 The scene takes place on a ship at sea.

0.1 *noise . . . lightning* Here as throughout
the text, the stage directions tend to be
descriptive, and seem designed for a
reader rather than for an acting company
preparing a production. It has been suggested that they were revised and am-
plified by Ralph Crane when he copied the
text for publication. (See Introduction,
The Text, pp. 57-8.) The usage is characteristically loose: Jacobean theatres had
lightning machines, and a *noise of thunder
and lightning heard* need not imply that no
visual effects accompanied the sound of
thunder.

3 **Good** Not good cheer, but either an expression of satisfaction at the boatswain's
presence ('good, you're here') or a
contraction of 'good fellow', as below,
ll. 15 and 19.
yarely quickly, smartly

4.1 *Enter Mariners* The Folio indicates no
exit for them, though they re-enter at

l. 50.1. Presumably they come and go
individually and severally throughout the
scene, and reappear as a group only when
the ship is about to founder.

5 **Cheerly** F's 'cheerely', perhaps trisyllabic
and equivalent to the modern 'cheerily'

6-8 **Take . . . enough** Given the confusions
of verse and prose in Crane manuscripts,
it is possible that this passage was intended as two blank verse lines.

6 **Tend** pay attention

7 **Blow . . . wind** i.e. do your worst

7-8 **if room enough** so long as there is
enough open sea, without reefs or rocks,
for the ship to ride out the storm

8.1 *Alonso* a variant of Alphonso, which is
the normal English form. For a discussion
of possible historical models for Alonso,
Ferdinand and Prospero, see the Introduction, pp. 42-3.
Sebastian, Antonio previously coupled as
the names of a shipwreck victim and his
adoring benefactor in *Twelfth Night*

(*To the Mariners*) Play the men.   10

BOATSWAIN I pray now, keep below.

ANTONIO Where is the master, bos'n?

BOATSWAIN Do you not hear him? You mar our labour.
Keep your cabins—you do assist the storm.

GONZALO Nay, good, be patient.

BOATSWAIN When the sea is. Hence! What cares these
roarers for the name of king? To cabin; silence! Trouble
us not.

GONZALO Good, yet remember whom thou hast aboard.

BOATSWAIN None that I love more than myself. You are a   20
councillor; if you can command these elements to silence,
and work the peace of the present, we will not hand a rope
more—use your authority. If you cannot, give thanks
you have lived so long, and make yourself ready in your
cabin for the mischance of the hour, if it so hap. (*To the
Mariners*)—Cheerly, good hearts! (*To the courtiers*)—Out
of our way, I say!   *Exit*

GONZALO I have great comfort from this fellow. Methinks he
hath no drowning mark upon him—his complexion is
perfect gallows. Stand fast, good Fate, to his hanging,   30
make the rope of his destiny our cable, for our own doth

---

12 bos'n] F (boson)   21 councillor] WILSON; counsellor F   22 present] F; presence KERMODE
(*conj.* Maxwell)   25–6 *To . . . Mariners*] This edition; *not in* F   26 *To . . . courtiers*] This
edition; *not in* F

10 **Play the men** Act like men: the remark is
officious and condescending, but Alonso
is the king.

16 **cares** For the form, see Abbott, 333.

17 **roarers** roaring winds and waves, with
an overtone of rioters

20 **None . . . myself** 'I am nearest to myself'
was proverbial (Tilley N57). Kittredge
quotes Terence: 'Proximus sum egomet
mihi' (*Andria* 4.1.12).

21 **councillor** The Folio spelling, counsellor,
implies not only a member of the Privy
Council, but an adviser and persuader.

22 **work the peace of the present** Kermode,
following a conjecture of J. C. Maxwell,
emends 'present' to 'presence'—i.e. the
royal presence—and sees in the phrase a

court metaphor developing from 'You are
a councillor . . .'. But, as Kermode
acknowledges, there is no parallel for the
expression, and the emendation is unnecessary and far-fetched: compare *Win-
ter's Tale* 4.1.13–14: 'make stale | The
glistering of the present.'
hand handle

26 **Cheerly** See l. 5 and note.

28–30 **he . . . gallows** The proverb was 'he
that is born to be hanged shall never be
drowned' (Tilley B139). Compare *Two
Gentlemen* 1.1.156–8.

29 **complexion** character, as indicated by
the physiognomy

31–2 **doth little advantage** is of little use to us

little advantage. If he be not born to be hanged, our
case is miserable.   *Exeunt*
    Enter Boatswain

BOATSWAIN Down with the topmast! Yare! Lower, lower!
Bring her to try with main-course. (*A cry within*) A
plague upon this howling! They are louder than the
weather or our office.
    Enter Sebastian, Antonio, and Gonzalo
Yet again? What do you here? Shall we give o'er and
drown? Have you a mind to sink?

SEBASTIAN A pox o' your throat, you bawling, blasphemous,   40
incharitable dog!

BOATSWAIN Work you, then.

ANTONIO Hang, cur, hang, you whoreson insolent
noisemaker! We are less afraid to be drowned than thou
art.

GONZALO I'll warrant him for drowning, though the ship
were no stronger than a nutshell and as leaky as an
unstanched wench.

---

33 *Exeunt*] THEOBALD; *Exit* F   35 *A cry within*] JOHNSON (*see next note*)   35–7.1 A plague
. . . *Gonzalo*] CAMBRIDGE; A plague—*A cry within. Enter . . . Gonzalo* F

34 **Down with the topmast** 'The topmast is
lowered to reduce the weight aloft and
check the drift toward the shore'
(Kittredge). For a detailed explanation of
the seamanship in this scene, see Appendix A.

35 **Bring . . . main-course** Kittredge explains,
'Heave to, under the mainsail. *Course* for
"sail" is common. The manœuvre is to
keep the ship close to the wind and away
from the island until they can weather it
and reach the open sea again.' To *try* the
ship was to lower the mainsail and keep as
close to the wind as possible.

35 **A plague** F follows these words with a
long dash. Kermode suggests that this
may indicate the deletion of a string of
oaths: the use of profanity in stage plays
was severely restricted by a statute of
1606 (Chambers, iv. 338). But expurgation seems an unlikely explanation,
since the play was written after 1606,
and Shakespeare would have known
what oaths were forbidden. The conjecture would be more plausible if the dash

came *before A plague*; the boatswain's sentence is complete, and it is difficult to see
how a string of oaths could have been
included between *plague* and *upon*. As the
text stands there is little to account for
Sebastian's charge that the Boatswain is
'bawling, blasphemous, incharitable'
(40–1).

36 **They** the passengers
37 **our office** we at our work
38 **give o'er** give up
40 **blasphemous** possibly referring to oaths
deleted from l. 35, though the word need
not imply impiety, but could mean simply
abusive, slanderous (*OED* 2)
46 **for** against: Gonzalo repeats his joke of
ll. 28–9.
47–8 **as an unstanched wench** Both E. A. M.
Colman (*Dramatic Use of Bawdy*) and Eric
Partridge (*Shakespeare's Bawdy*) take the
joke to be about menstruation without
the use of absorbent padding, but *unstanched* can mean unsatisfied, and *leaky* may
therefore instead imply sexual arousal.

BOATSWAIN Lay her a-hold, a-hold! Set her two courses off
to sea again; lay her off!   50
    Enter Mariners wet
MARINERS All lost! To prayers, to prayers! All lost! *Exeunt*
BOATSWAIN What, must our mouths be cold?
GONZALO

The King and Prince at prayers, let's assist them,
For our case is as theirs.
SEBASTIAN   I'm out of patience.
ANTONIO

We are merely cheated of our lives by drunkards.
This wide-chopped rascal—would thou mightst lie
    drowning
The washing of ten tides!   *Exit Boatswain*
GONZALO   He'll be hanged yet,
Though every drop of water swear against it,
And gape at wid'st to glut him.
    A confused noise within
    'Mercy on us!'—'We split, we split!'—'Farewell,   60
    my wife and children!'—'Farewell, brother!'
    —'We split! we split! we split!'
ANTONIO Let's all sink wi' th' King.
SEBASTIAN Let's take leave of him.   *Exit with Antonio*
GONZALO Now would I give a thousand furlongs of sea for
an acre of barren ground—long heath, brown furze, any-
thing. The wills above be done, but I would fain die a dry
death.   *Exit*

---

51 *Exeunt*] THEOBALD; *not in* F   54 I'm] F (I' am)   57 *Exit Boatswain*] DYCE; *not in* F
64 *Exit with Antonio*] CAMBRIDGE; *Exit* F   66 furze] F (firrs) (*see note*)

49 **Lay her a-hold** Bring the ship close to the
wind so as to hold it; to do this more sail
must be set, hence the order immediately
following. This is the only example of the
term in print, but A. F. Falconer points
out that it survives in New England boat-
ing parlance. See Appendix A.
    two courses the foresail and mainsail
(*course* = sail)
50 **lay her off** get her out to sea
52 **must our mouths be cold** To be cold in the
mouth, i.e. dead, was proverbial (Dent
M1260.1). Some editors interpret the line
to mean that the Boatswain here swigs a

drink, thereby providing some basis for
Antonio's charge of drunkenness at l. 55.
55 **merely** completely
56 **wide-chopped** big-mouthed
57 **The washing of ten tides** Antonio hyper-
bolically alludes to the punishment for
pirates, which was to be hanged 'at the
low water mark, and there to remain till
three tides had overflowed them' (Stow,
Survey of London (1599), p. 347).
59 **glut** swallow
66 **long heath, brown furze** heather and
gorse; the passage has suffered much
interpretation. Rowe first printed *furze* for

1.2   *The island.*
       Enter Prospero and Miranda

MIRANDA

If by your art, my dearest father, you have
Put the wild waters in this roar, allay them.
The sky, it seems, would pour down stinking pitch,
But that the sea, mounting to th' welkin's cheek,
Dashes the fire out. O, I have suffered
With those that I saw suffer: a brave vessel—
Who had, no doubt, some noble creature in her—
Dashed all to pieces! O, the cry did knock
Against my very heart—poor souls, they perished.
Had I been any god of power, I would   10

the Folio's 'firrs'—this is not an emendation, but a spelling variant (the plural appears in the Folio as 'firzes' at 4.1.180). Hanmer, because *furze* is not brown, inserted a comma and emended to 'broom'; this has been accepted by many editors, including Kermode. Hanmer also changed *long heath* to 'ling, heath': this has generally been rejected because *long heath* (like 'small heath') is the name of a particular plant (Lyte's *Herbal* (1576), pp. 677–8). But *brown* also requires no emendation. Gonzalo says he will be satisfied with an acre of anything, even the most wild and parched vegetation. As for 'firrs', it is possible that Shakespeare intended the tree, not the shrub: Captain John Smith says of Virginia that 'the rocky cliffs' are 'overgrown with Firre' (cited in *OED*, s.v.). Only Dover Wilson prints 'firs'.

1.2.0.1 **Prospero** The name means 'fortunate' or 'prosperous' (literally 'according to one's hopes'). Jonson used it and Stephano in *Every Man in his Humour* (1601), in which, according to the cast list in the Jonson folio, Shakespeare performed. Possible historical sources for the name are discussed in the Introduction, pp. 42–3.

Miranda literally 'wonderful', 'to be wondered at'

1   **art** used throughout to refer to Prospero's magic powers, and in the Folio text capitalized throughout. The capitalization, however, is apparently an idiosyncrasy of Ralph Crane's: the word is invariably capitalized in Crane transcriptions,

whether it refers to magic or not. The operative meanings include 'learning', 'science', most especially 'skill in applying the principles of a special science' (*OED* 4). For magic as an art, compare *Winter's Tale* 5.3.110–11: 'If this be magic, let it be an art | Lawful as eating.'

3   **pitch** implying chiefly its smell and blackness here, but also with moral overtones ('pitch defiles') and possibly an ironic ambiguity as well: its practical use was for caulking ships.

4   **welkin** The word originally meant either cloud or sky. By Shakespeare's time the cloud meaning had dropped out, and the word was, in southern English, exclusively literary.
    cheek common in personifications of both heaven and the sea. The 'fire' of l. 5 suggested to Dover Wilson that *OED*'s 'side pieces of a grate or stove' was also applicable, but this seems to be a later usage. Kermode compares *Othello* 4.2.75, 'forges of my cheeks', but the image there refers to the bellows of the forge, not to its structure.

5   **fire** the lightning, imagined as boiling the pitch of l. 3

6   **brave** fine, noble

7   **Who** 'Especially used after antecedents that are lifeless or irrational when personification is employed' (Abbott 259.2b).

10 **god of power** The power is both Prospero's magic generally, and, specifically, the raising of storms, as in George Herbert's 'The Bag' (*c.* 1630), ll. 5 and 9: 'Storms are the triumph of his art . . . The God of Power, as he did ride . . .'

Have sunk the sea within the earth or ere
It should the good ship so have swallowed, and
The fraughting souls within her.

PROSPERO   Be collected.
No more amazement. Tell your piteous heart
There's no harm done.

MIRANDA   O, woe the day!

PROSPERO   No harm.
I have done nothing but in care of thee,
Of thee, my dear one, thee, my daughter, who
Art ignorant of what thou art; naught knowing
Of whence I am, nor that I am more better
Than Prospero, master of a full poor cell,   20
And thy no greater father.

MIRANDA   More to know
Did never meddle with my thoughts.

PROSPERO   'Tis time
I should inform thee farther. Lend thy hand
And pluck my magic garment from me.
         Miranda helps him to disrobe

1.2.24.1 *Miranda . . . disrobe*] This edition; *Lays down his mantle* POPE; *not in* F

---

11 **or ere** The two words are cognate, both meaning 'before'. The doubling is for emphasis. See Abbott 131.

13 **fraughting** 'that forms freight or cargo' (*OED*); not normally used of people, but see *Troilus* Prologue 4 and 13: 'their ships | Fraught with the ministers and instruments | Of cruel war'; 'the deep-drawing barks do there disgorge | Their warlike fraughtage . . .'

14 **amazement** both overwhelming fear (*OED* 3) and overwhelming wonder (*OED* 4), comprising, with the *piteous heart* immediately following, the full Aristotelian response to tragedy.
   piteous here, feeling pity; Shakespeare also uses the word to mean 'pitiful', e.g. *Titus* 5.1.66.

19 **more better** higher in rank. For the double comparative, see Abbott 11.

20 **cell** technically a single-chamber dwelling, often with monastic implications; by the late-sixteenth century used poetically for 'a small and humble dwelling, a cottage' (*OED*); not applied to prisons until the eighteenth century.

22 **meddle with** The original meaning is 'mix with' with a sexual connotation persisting until well into the seventeenth century. The modern pejorative usage, 'interfere with', appears to be the most common one by Shakespeare's time. It is, therefore, worth noting both the passiveness Miranda claims for her thoughts here, and the clear contradiction of that claim in her recollection of her frequent 'bootless inquisition', l. 35.

24–5 **magic garment . . . Lie there, my art.** Prospero refers to his cloak of office. Steevens cited Fuller on Lord Burleigh: 'At night when he put off his gown, he used to say, Lie there, *Lord Treasurer*. . .' (*The Holy State* (Cambridge, 1642), p. 269).

                                           So.
Lie there, my art.—Wipe thou thine eyes; have
    comfort.
The direful spectacle of the wreck, which touched
The very virtue of compassion in thee,
I have with such provision in mine art
So safely ordered that there is no soul,
No, not so much perdition as an hair   30
Betid to any creature in the vessel
Which thou heard'st cry, which thou saw'st sink. Sit
    down,
For thou must now know farther.
        They sit
MIRANDA   You have often
Begun to tell me what I am, but stopped,
And left me to a bootless inquisition,
Concluding, 'Stay, not yet.'
PROSPERO   The hour's now come;
The very minute bids thee ope thine ear.
Obey, and be attentive. Canst thou remember
A time before we came unto this cell?
I do not think thou canst, for then thou wast not   40
Out three years old.
MIRANDA   Certainly, sir, I can.
PROSPERO
By what? By any other house or person?

33.1 *They sit*] This edition; *not in* F

26 **spectacle** The predominant meaning is
   'theatrical display or pageant', 'safely
   ordered' by Prospero as presenter, l. 29.
29 **no soul** an anacoluthon. The omitted
   verb, 'perished', is implied in 'perdition',
   l. 30. Editors since Steevens have inserted
   a dash after 'soul' to indicate a break in
   Prospero's thought, but the sentence
   reads smoothly, and the syntax is not
   unusual, especially in the late plays.
30 **not . . . hair** 'Not to hurt (or lose) a hair'
   was proverbial (see Dent H26.1). Ariel
   uses the same locution at 1.2.217.
   perdition loss; the word is characterized
   by *OED* as 'affected or rhetorical' usage,
   and appears in *Henry V* 3.6.100 as a

comic pomposity of Fluellen's. Ariel uses
the word again on Prospero's instructions, 3.3.77.
32 **Which . . . which** whom . . . which. See
   Abbott, 265.
35 **bootless** unsuccessful
   inquisition a formal or legal process of inquiry; like 'perdition', l. 30, rhetorical
   usage
38 **Obey, and be** F's comma after *ear* (l. 37)
   may imply that these are not imperatives
   but, like *ope*, infinitives dependent on *bids*.
41 **Out** beyond, hence 'fully'
42 **By what** i.e. by what image (in your
   memory)

Of anything the image tell me that
Hath kept with thy remembrance.

MIRANDA   'Tis far off,
And rather like a dream than an assurance
That my remembrance warrants. Had I not
Four or five women once that tended me?

PROSPERO
Thou hadst, and more, Miranda; but how is it
That this lives in thy mind? What seest thou else
In the dark backward and abyss of time?   50
If thou rememb'rest aught ere thou cam'st here,
How thou cam'st here thou mayst.

MIRANDA   But that I do not.

PROSPERO
Twelve year since, Miranda, twelve year since,
Thy father was the Duke of Milan, and
A prince of power—

MIRANDA   Sir, are you not my father?

PROSPERO
Thy mother was a piece of virtue, and
She said thou wast my daughter; and thy father
Was Duke of Milan, and his only heir
And princess no worse issued.

MIRANDA   O, the heavens!
What foul play had we that we came from thence?   60
Or blessèd was't we did?

---

50 abyss] F (abysme)   59 princess no] KNIGHT; princesse; no F

43 **Of . . . me** describe to me whatever: the memory is assumed to be visual.

46 **warrants** guarantees as true (*OED* 4)

50 **backward** 'the past portion (of time)' (*OED* C2, citing only this passage and an 1870 example which is clearly alluding to it). Abbott 77 compares 'I was an inward of his', *Measure* 3.2.138.

53 **year** often used as a plural: see *OED*'s citations s.v. 1d. Many editors argue that the word is a disyllable on its first appearance and a monosyllable on its second. This is conceivable, but there is no reason to assume that the line must be metrically regular.

54 **Milan** accented on the first syllable

56 **Thy mother** See Introduction, pp. 16, 18.
piece prototype, model

59 **And princess** Since Pope, often emended to 'a princess'. Dyce defended the change, citing several instances where 'and' has clearly been substituted for 'a' in Shakespearian texts. A compositor's error may be involved; but the Folio as it stands is perfectly coherent, and the presumed awkwardness of the line is no ground for emendation.
no worse issued was no less nobly descended

PROSPERO   Both, both, my girl.
By foul play, as thou sayst, were we heaved thence,
But blessedly holp hither.
MIRANDA   O, my heart bleeds
To think o'th' teen that I have turned you to,
Which is from my remembrance. Please you, farther.
PROSPERO
My brother, and thy uncle, called Antonio—
I pray thee mark me, that a brother should
Be so perfidious—he whom next thyself
Of all the world I loved, and to him put
The manage of my state, as at that time   70
Through all the signories it was the first,
And Prospero the prime duke, being so reputed
In dignity, and for the liberal arts
Without a parallel; those being all my study,
The government I cast upon my brother,
And to my state grew stranger, being transported
And rapt in secret studies. Thy false uncle—
Dost thou attend me?
MIRANDA   Sir, most heedfully.
PROSPERO
Being once perfected how to grant suits,
How to deny them, who t'advance, and who   80
To trash for overtopping, new created

77 studies. Thy] F4; studies, thy F1

63 **holp** helped (shortened form of the old p.p. *holpen*)
64 **teen** trouble
65 **from** away from, not present in
68–74 **he . . . parallel** Most editors repunctuate in an attempt to give some sense of order to the syntax, but the run-on clauses are dramatically coherent, expressing Prospero's excitement and retrospective rage.
70 **manage** administration
71 **signories** both lordships and domains, specifically applied to the Italian city-states
73 **liberal arts** technically those 'considered "worthy of a free man"; opposed to *servile* or *mechanical* . . . [arts] suitable to persons of superior social station' (*OED* s.

liberal). Specifically, grammar, logic, and rhetoric (the *trivium*), and arithmetic, geometry, music, and astronomy (the *quadrivium*)
76 **state** the dukedom—either the office or the country
76–7 **transported | And rapt** Both words literally mean 'physically carried away': Prospero describes his studies as a prefiguration of his abduction and dispatch to the island.
79 **perfected** completely versed in, coming to a mastery of (the word is accented on the first and third syllables)
81 **trash** 'to check (a hound) by a cord or leash' (*OED*)
overtopping gaining too much power or authority

The creatures that were mine, I say: or changed 'em,
Or else new formed 'em; having both the key
Of officer and office, set all hearts i'th' state
To what tune pleased his ear, that now he was
The ivy which had hid my princely trunk,
And sucked my verdure out on't—thou attend'st not!

MIRANDA

O, good sir, I do!

PROSPERO   I pray thee mark me:
I thus neglecting worldly ends, all dedicated
To closeness and the bettering of my mind   90
With that which, but by being so retired,
O'er-prized all popular rate, in my false brother
Awaked an evil nature, and my trust,
Like a good parent, did beget of him
A falsehood in its contrary as great
As my trust was, which had, indeed, no limit,
A confidence sans bound. He being thus lorded,
Not only with what my revenue yielded,
But what my power might else exact, like one

82 **creatures** dependants (whose offices have been *created*)
82–3 **or changed 'em, | Or else new formed 'em** either changed (the allegiance and/or the duties of) existing officials, or else created new ones. *Changed* is equivalent to *new created* in l. 81, and in contrast to *new formed* in l. 83.
83–4 **both . . . office** control over both officials and administration
83 **key** The keys of his office become, with 'set . . . to what tune', the keys of musical notation.
85 **that** so that
86–7 **ivy . . . on't** a familiar topos, usually representing the perils of symbiotic relationships: compare the February Eclogue of Spenser's *Shepherd's Calendar*. Geoffrey Whitney's *Choice of Emblems* (1586) uses an elm and a grapevine to illustrate the advantages of the arrangement; the motto is '*Amicitia, etiam post mortem durans*' [friendship lasting even after death] (p. 62).
87 **verdure** sap, vitality, hence power

90 **closeness** privacy
91 **but** merely
92 **O'er-prized all popular rate** exceeded the common people's understanding (*o'er-prized* = were priced beyond the reach of). The point is that what was incomprehensible and disruptive was the fact of retirement and secrecy, and the elitism these implied, not anything inherently mysterious about the studies themselves (they were, after all, 'the liberal arts').
94 **Like a good parent** 'Alluding to the observation that a father above the common rate commonly has a son below it' (Johnson). Erasmus in the *Adagia* expounds 'Heroum filii noxae' a Latin version of *Odyssey*, ii. 315, cited by Johnson in support of his observation. And Tilley records the proverb 'Great men's sons seldom do well' (M611).
97 **sans** without, a common loan-word at this time
lorded turned into a lord
98 **revenue** pronounced revènue

Who, having into truth by telling of it,   100
Made such a sinner of his memory
To credit his own lie, he did believe
He was indeed the duke, out o'th' substitution
And executing th'outward face of royalty
With all prerogative. Hence his ambition growing—
Dost thou hear?
MIRANDA   Your tale, sir, would cure deafness.
PROSPERO
To have no screen between this part he played
And him he played it for, he needs will be
Absolute Milan. Me, poor man, my library
Was dukedom large enough. Of temporal royalties   110
He thinks me now incapable; confederates—
So dry he was for sway—with' King of Naples
To give him annual tribute, do him homage,
Subject his coronet to his crown, and bend
The dukedom yet unbowed—alas, poor Milan!—
To most ignoble stooping.
MIRANDA   O, the heavens!
PROSPERO Mark his condition, and th'event; then tell me
If this might be a brother.
MIRANDA   I should sin

100–2 **Who . . . lie** The syntax is, 'who, having made of his memory such a sinner against (*into*) truth as to credit his own lie by telling it'.
103 **out o'th' substitution** as a consequence of having taken my place
107–8 **To have no screen . . . for** to have no barrier between his role and himself, to act for himself. The metaphor is confusing because it in fact characterizes Prospero's situation, not Antonio's: it is Prospero who set up Antonio as a screen between himself and his office.
108–11 **will be . . . thinks . . . confederates** As Prospero relives the experience, his tenses change from past to present and future.
109 **Absolute** 'free from all external restraint or interference; unrestricted, unlimited, independent' (*OED* iii. 7)
Me for me, or as for me
110 **temporal royalties** the prerogatives of rule, as opposed to the spiritual prerogatives afforded by his intellectual pursuits

111 **confederates** conspires: clearly pejorative here, though generally not so in the period
112 **dry** thirsty, hence eager. Compare *Troilus* 2.3.234 (Nestor of Ajax): 'his ambition is dry.' The metaphor is to be dramatically realized in the drunken conspiracy of Caliban, Stephano, and Trinculo.
115 **yet hitherto**
117 **his condition, and th'event** the terms of his agreement with Naples and its outcome
118–20 **I should sin . . . bad sons** Miranda takes Prospero's attack on Antonio to imply an accusation of adultery against Prospero's mother. The underlying assumptions here have almost the status of a topos in the period: compare the joking in *Much Ado* 1.1.104–6, another play with very ambivalent attitudes towards women. See the Introduction, pp. 16–17.

To think but nobly of my grandmother:
Good wombs have borne bad sons.

PROSPERO   Now the condition.   120
This King of Naples, being an enemy
To me inveterate, hearkens my brother's suit,
Which was that he, in lieu o'th' premises
Of homage and I know not how much tribute,
Should presently extirpate me and mine
Out of the dukedom, and confer fair Milan,
With all the honours, on my brother; whereon,
A treacherous army levied, one midnight
Fated to th' purpose did Antonio open
The gates of Milan, and i'th' dead of darkness   130
The ministers for th' purpose hurried thence
Me and thy crying self.

MIRANDA   Alack, for pity!
I not rememb'ring how I cried out then
Will cry it o'er again— it is a hint
That wrings mine eyes to't.

PROSPERO   Hear a little further,
And then I'll bring thee to the present business
Which now's upon's; without the which this story
Were most impertinent.

MIRANDA   Wherefore did they not
That hour destroy us?

PROSPERO   Well demanded, wench:

---

123 **in lieu o'th' premises** in return for the
conditions agreed upon
125 **presently** immediately
    extirpate literally, 'uproot'; the word
    could mean either exterminate or drive
    off.
128 **levied, one midnight** The rhythm is am-
    biguous. Shakespeare usually accents
    midnight on the first syllable, which would
    suggest a trisyllabic *levièd*, but the
    alternative accentuation occasionally
    occurs. Either way, the metre requires an
    obsolete accentuation at some point.
129 **Fated** appointed by fate; Prospero
    subsequently claims that 'bountiful
    Fortune' is 'Now my dear lady',

ll. 178–9.
131 **ministers** agents
134 **hint** occasion (literally 'something one
    seizes on'). Compare 2.1.3.
135 **wrings** The metaphorical use is rare
    without 'tears' as its object.
    to't for the purpose (Abbott 186)
137 **the which** referring to the previous
    which (Abbott 270)
138 **impertinent** 'not pertaining to the
    matter at hand' (*OED* 2)
139 **wench** originally a young woman or
    girl child; also, in Shakespeare's time, 'a
    familiar or endearing form of address;
    used chiefly in addressing a daughter,
    wife or sweetheart' (*OED* 1c)

My tale provokes that question. Dear, they durst not,   140
So dear the love my people bore me, nor set
A mark so bloody on the business; but
With colours fairer painted their foul ends.
In few, they hurried us aboard a barque,
Bore us some leagues to sea, where they prepared
A rotten carcase of a butt, not rigged,
Nor tackle, sail, nor mast—the very rats
Instinctively have quit it. There they hoist us
To cry to th' sea that roared to us, to sigh
To th' winds, whose pity, sighing back again,   150
Did us but loving wrong.

MIRANDA   Alack, what trouble
Was I then to you!

PROSPERO   O, a cherubin
Thou wast that did preserve me. Thou didst smile,
Infusèd with a fortitude from heaven,
When I have decked the sea with drops full salt,
Under my burden groaned, which raised in me

140–1 **Dear ... So dear** Hanmer omitted the first *dear* and Staunton conjectured that the second was a printer's error for 'dare'. Kermode claims that 'this repetition is not a deliberate chiming', but concludes that there is no good reason to emend it. Since the repetition relates the love of parent and child to the love of prince and people, there is every reason to leave it as it is.

142 **A mark so bloody** 'Those who came in at the death were marked with the blood shed by the deer' (Kermode).

144 **few** i.e. few words

a barque In fact, Milan is not a port.

146 **carcase** 'The decaying skeleton of a vessel' (*OED* 5)

butt literally a barrel or tub; but (unlike tub) not recorded as a slang term for a boat. The word is apparently etymologically unrelated to Italian *botto*, a kind of sloop, or to French *boute*, a leathern vessel; but Shakespeare may be using it in the belief that it is.

148 **hoist** invariably treated as the past tense (see Abbott 342), but, like *have quit* immediately preceding, it may be another of

Prospero's characteristic shifts into present tense. This sort of temporal inconsistency is not uncommon in the period (compare 'Fair stood the wind for France | When we our sails advance'), but in this play it is employed only by Prospero.

152 **cherubin** originally a plural, but used as the normal singular in English until the seventeenth century. *Cherub* began to be used, at first only in Biblical translations, after Wyclif (d. 1384).

155 **decked** adorned; generally glossed 'strewn', but *OED* gives no support for this. Sisson observes that the whole of Prospero's language here is elaborate and conceited, with a good deal of irony (*New Readings*, i. 44).

156–7 **Under my burden ... An undergoing stomach** *Undergoing stomach* = courage to endure; *stomach* implies variously the inmost part, temper, especially 'spirit, courage, valour, bravery' (*OED* 8). But also, the phrase contributes to a birth metaphor begun in l. 155: *burden* is the contents of the womb (*OED* 4), and a *groaning* is a lying in (*OED* 2).

An undergoing stomach to bear up
Against what should ensue.

MIRANDA   How came we ashore?

PROSPERO By providence divine;
Some food we had, and some fresh water, that   160
A noble Neapolitan, Gonzalo,
Out of his charity, who being then appointed
Master of this design, did give us, with
Rich garments, linens, stuffs, and necessaries,
Which since have steaded much; so of his gentleness,
Knowing I loved my books, he furnished me
From mine own library with volumes that
I prize above my dukedom.

MIRANDA   Would I might
But ever see that man!

PROSPERO (*rising*)   Now I arise.
Sit still, and hear the last of our sea-sorrow: 170
Here in this island we arrived, and here
Have I, thy schoolmaster, made thee more profit
Than other princes can that have more time
For vainer hours, and tutors not so careful.

MIRANDA
Heavens thank you for't. And now I pray you, sir,

---

159 divine;] ROWE; divine, F   169 *rising*] This edition; *not in* F   173 princes] F (princesse)
(*see note*) 175 Heavens] *the variant reading Heven has been shown to be a ghost:* ShQ 26 (1975),
pp. 213–14

159 **By providence divine** Dover Wilson
assumed that the short line implied a cut
here, but rhetorical and dramatic effec-
tiveness seem sufficient explanations for
the metrical irregularity. (Compare, for
example, 1.2.235 and 253.) Most editors
since Pope (including Kermode) replace
F's comma with a period. The syntax is
ambiguous; in F's punctuation,
providence governs not only their coming
ashore but Gonzalo's charitable interven-
tion as well. There is no way for modern
English to preserve this ambiguity, but a
less strong mark of punctuation is war-
ranted.
165 **steaded** been useful or advantageous
   gentleness both kindness and nobility
169 **Now I arise** both literally, as Prospero

prepares to exercise his control over the
shipwreck victims, and figuratively, as he
sees his fortunes turn (compare 'my
zenith . . .', l. 181). There is no clear indication of when he resumes his magic
cloak, but 'I am ready now' (l. 187)
provides an appropriate point.
170 **Sit still** remain seated
172 **profit** the verb, not the noun
173 **princes** F's 'princesse' is a characteristic
   spelling of *princes* in Ralph Crane
   manuscripts. (*OED* does not record it, but
   compare *guesse* for *guests*, which is recorded.) The word is a generic term for
   royal children of either sex. See Roberts,
   pp. 228–9.
174 **careful** both caring and taking trouble

For still 'tis beating in my mind, your reason
For raising this sea-storm.

PROSPERO   Know thus far forth:
By accident most strange, bountiful Fortune,
Now my dear lady, hath mine enemies
Brought to this shore; and by my prescience   180
I find my zenith doth depend upon
A most auspicious star, whose influence
If now I court not, but omit, my fortunes
Will ever after droop. Here cease more questions:
Thou art inclined to sleep. 'Tis a good dulness,
And give it way—I know thou canst not choose.
          Miranda sleeps
(*Calling*) Come away, servant, come.
          ⌈*Puts on his cloak*⌉   I am ready now.
Approach, my Ariel. Come.
          Enter Ariel

ARIEL

All hail, great master, grave sir, hail! I come
To answer thy best pleasure, be't to fly,   190
To swim, to dive into the fire, to ride
On the curled clouds; to thy strong bidding task
Ariel and all his quality.

PROSPERO   Hast thou, spirit,

---

186. I *Miranda sleeps*] THEOBALD; *not in* F   187 *Calling*] This edition; *not in* F   *Puts on his cloak*] This edition; *not in* F

178–9 **Fortune, | Now my dear lady** Fortuna was characteristically fickle.

181 **zenith** technically the highest point of the celestial sphere, and also, here, the top of Fortune's wheel; hence, the culmination of Prospero's good fortune

182 **influence** astrological powers

183 **omit** disregard, 'fail or forbear to use' (*OED*)

185 **dulness** drowsiness

187 **I am** The metre suggests that Crane may be expanding 'I'm' here.

188.1 **Ariel** literally 'lion of God', used by Isaiah as an epithet for Jerusalem (29: 1). The name appears as a spirit in many magical texts, especially Agrippa's *De Occulta Philosophia*, 3.28.436 and 3.24.416; but in these cases it is invariably a spirit of earth, not air. It is also the name of an evil angel, formerly a pagan god. (See W. Stacy Johnson, 'The Genesis of Ariel', *ShQ*, 2 (1951), pp. 205–11; and, for other references, Robert H. West, 'King Hamlet's Ambiguous Ghost', *PMLA* 70 (1955), p. 1112, n. 13.)

190–2 **be't . . . clouds** Ariel declares himself at home in the fluid and volatile elements; Prospero adds 'earth' at l. 255.

193 **quality** either fraternity (*OED* 5), i.e. the other spirits, or abilities (*OED* 2b)

Performed to point the tempest that I bade thee?
ARIEL   To every article.
I boarded the King's ship; now on the beak,
Now in the waist, the deck, in every cabin,
I flamed amazement. Sometime I'd divide
And burn in many places; on the topmast,
The yards and bowsprit would I flame distinctly,   200
Then meet and join. Jove's lightning, the precursors
O'th' dreadful thunder-claps, more momentary
And sight-outrunning were not; the fire and cracks
Of sulphurous roaring the most mighty Neptune
Seem to besiege and make his bold waves tremble,
Yea, his dread trident shake.

PROSPERO   My brave spirit!
Who was so firm, so constant, that this coil
Would not infect his reason?
ARIEL   Not a soul
But felt a fever of the mad, and played
Some tricks of desperation. All but mariners   210
Plunged in the foaming brine and quit the vessel,

211–12 vessel . . . me:] ROWE: vessell; | Then all a fire with me F

194 **Performed to point** presented in exact
detail
195 **article** The metaphor is of a legal docu-
ment.
196 **beak** prow
197 **in the waist** amidships
deck 'In early craft there was a deck only
at the stern, so that sixteenth-century
writers sometimes use *deck* as equivalent
to *poop*' (OED 2).
198 **flamed amazement** appeared as flame,
producing terror. Here and at l. 200, most
editors refer to St Elmo's fire and cite
various travel narratives, in which, however, the phenomenon is generally treated
as a comforting omen: Hakluyt says that St
Elmo is 'the advocate of sailors'. Here the
manifestation produces 'desperation'; (l.
210). Cotgrave (1632) has, under *furole*,
'A little blaze of fire appearing by night on
the . . . sailyards, where it whirls and leaps
in a moment from one place to another;
some mariners call it St Hermes' fire; if it
come double 'tis held a sign of good luck, if
single, otherwise.' (Both cited in the
Variorum.) Strachey reports that the

phenomenon appeared during the Bermuda voyage, where it was observed 'with
much wonder and carefulness', but that
the observers were uncertain of its significance. (See Appendix B.) See also
Richard Eden's translation of Antonio
Pigafetta's account of Magellan's voyages,
where the mysterious fire is described two
pages before the name Setebos appears:
The Decades of the New World of West India
(London, 1555), pp. 217–18 (the reprint
in Eden's *History of Travel* (1577) is cited by
Kermode).
204 **sulphurous** Sulphur was popularly
associated with thunder and lightning,
from its use in explosives.
207 **coil** tumult, confusion
209 **a fever of the mad** such a fever as the
mad feel
211–12 **vessel . . . Ferdinand** Since Rowe's
time, only Dover Wilson has defended F's
punctuation. The question is whether
only the ship, or Ferdinand too, is 'all afire
with me'. Ariel's account immediately
preceding seems to limit his inflammatory
activity to the ship.

Then all afire with me: the King's son Ferdinand,
With hair up-staring—then like reeds, not hair—
Was the first man that leapt, cried 'Hell is empty,
And all the devils are here.'
PROSPERO   Why, that's my spirit.
But was not this nigh shore?
ARIEL   Close by, my master.
PROSPERO
But are they, Ariel, safe?
ARIEL   Not a hair perished.
On their sustaining garments not a blemish,
But fresher than before; and as thou bad'st me,
In troops I have dispersed them 'bout the isle.   220
The King's son have I landed by himself,
Whom I left cooling of the air with sighs
In an odd angle of the isle, and sitting,
His arms in this sad knot.
PROSPERO   Of the King's ship
The mariners say how thou hast disposed,
And all the rest o'th' fleet.
ARIEL   Safely in harbour
Is the King's ship, in the deep nook where once
Thou called'st me up at midnight to fetch dew
From the still-vexed Bermudas, there she's hid;
The mariners all under hatches stowed,   230

213 **up-staring** standing on end
214 **Hell is empty** Similar proverbial expressions are cited in Tilley (H403).
215 **devils** often monosyllabic, and possibly so here
217 **Not a hair perished** Ariel repeats Prospero's reassurance to Miranda: see 1.2.30.
218 **sustaining garments** They were buoyed up by their clothing, either because of the magical quality of the wreck, or naturally and briefly like Ophelia (*Hamlet* 4.7.175–83), but in this case long enough to enable them to reach the nearby shore.
223 **angle** corner
224 **this sad knot** Ariel folds his arms, implying sorrow. Compare *Titus* 3.2.4: 'Mar-

cus, unknit that sorrow-wreathen knot,' and *Julius Caesar* 2.1.240: 'Musing and sighing with your arms a-cross.'
228 **midnight . . . dew** the appropriate time and a common substance for the performance of magic. Caliban credits Sycorax with the use of 'wicked dew' at l. 321.
229 **still-vexed** always troubled by storms
Bermudas the only reference to Bermuda in the play. Kittredge cites a passage from Middleton and Webster, *Anything for a Quiet Life*, 5.1.355–7, in which the Bermudas are associated with both tempests and magic: 'The place I speak of [the Bermudas] has been kept with thunder, | With frightful lightnings, amazing noises, | But now, th'enchantment broke, 'tis the land of peace . . .'

Who, with a charm joined to their suffered labour,
I have left asleep; and for the rest o'th' fleet,
Which I dispersed, they all have met again,
And are upon the Mediterranean float,
Bound sadly home for Naples,
Supposing that they saw the King's ship wrecked,
And his great person perish.

PROSPERO   Ariel, thy charge
Exactly is performed; but there's more work.
What is the time o'th' day?

ARIEL   Past the mid-season.

PROSPERO

At least two glasses. The time 'twixt six and now   240
Must by us both be spent most preciously.

ARIEL

Is there more toil? Since thou dost give me pains,
Let me remember thee what thou hast promised,
Which is not yet performed me.

PROSPERO   How now? Moody?
What is't thou canst demand?

ARIEL   My liberty.

PROSPERO

Before the time be out? No more.

ARIEL   I prithee,
Remember I have done thee worthy service,
Told thee no lies, made no mistakings, served
Without or grudge or grumblings. Thou did promise
To bate me a full year.

PROSPERO   Dost thou forget   250
From what a torment I did free thee?

---

248 made no] ROWE; made thee no F

---

231 **Who** for 'whom' (Abbott 274)
    their suffered labour the toil they have
    undergone
234 **float** sea (*OED* 3)
240 **two glasses** i.e. two hours past noon.
    The reference here and at 5.1.223 is to
    hour glasses, not to the half-hour glasses
    used by mariners.
242 **pains** tasks: the complaint anticipates
    Caliban's charges against Prospero.

243 **remember** remind
244 **me** dative; see Abbott 220.
    Moody the first indication of Ariel's
    characteristic rebelliousness. Douce ob-
    served that 'the spirits or familiars attend-
    ing on magicians were always impatient
    of confinement' (citing Scot's *Discovery of
    Witchcraft* (1665), p. 228).
250 **bate me** deduct from my time

ARIEL   No.

PROSPERO

Thou dost, and think'st it much to tread the ooze
Of the salt deep,
To run upon the sharp wind of the north,
To do me business in the veins o'th' earth
When it is baked with frost.

ARIEL   I do not, sir.

PROSPERO

Thou liest, malignant thing! Hast thou forgot
The foul witch Sycorax, who with age and envy
Was grown into a hoop? Hast thou forgot her?

ARIEL

No, sir.

PROSPERO Thou hast. Where was she born? Speak; tell
       me.   260

ARIEL

Sir, in Algiers.

PROSPERO   O, was she so—I must
Once in a month recount what thou hast been,
Which thou forget'st. This damned witch Sycorax,
For mischiefs manifold and sorceries terrible
To enter human hearing, from Algiers

261, 265 Algiers] F (Argier)   265 human] F (humane)

255 **veins** either mineral deposits ('veins of
ore') or the channels of underground
streams (*OED* s. vein ii. 6)
256 **baked** The operative meaning of bake is
'to harden as frost does' (*OED* s.v. 3,
citing Barnabe Googe, 1586). Shakespeare does not use the verb in this way
elsewhere.
258 **Sycorax** The name has never been
satisfactorily explained. It is usually
etymologized from the Greek *sus* (pig) and
korax (raven); only the latter of these
seems right. The figure is largely derived
from Ovid's account of Medea in *Metamorphoses* 7, and the name sounds like an
epithet for Medea, the Scythian raven.
(See the Introduction, pp. 19–20.) Overtones of Circe may also be present: Kermode points out that according to the

mythographers she 'was born in Colchis,
the district of the Coraxi tribe'—Medea
also came from Colchis. The account of
Sycorax's career is presented as deriving
from Prospero's memory, but in fact the
memory is Ariel's. Prospero never saw
Sycorax, who died before he came to the
island.
261 **O, was she so** Dover Wilson thought 'P.
about to contradict A. but does not do so;
and the text leaves us in doubt as to the
birthplace of Sycorax'. Since the memory
is Ariel's to begin with, this can hardly be
intended. Kermode admitted Dover Wilson's speculation, but suggested sarcasm
as a more likely possibility. Some sort of
sarcastic emphasis is clearly being expressed—'So you do remember after all!'

I.2   *The Tempest*

Thou know'st was banished—for one thing she did
They would not take her life. Is not this true?
ARIEL Ay, sir.
PROSPERO
This blue-eyed hag was hither brought with child,
And here was left by th' sailors. Thou, my slave,   270
As thou report'st thyself, was then her servant,
And for thou wast a spirit too delicate
To act her earthy and abhorred commands,
Refusing her grand hests, she did confine thee,
By help of her more potent ministers
And in her most unmitigable rage,
Into a cloven pine, within which rift
Imprisoned thou didst painfully remain
A dozen years; within which space she died
And left thee there, where thou didst vent thy groans   280
As fast as mill-wheels strike. Then was this island—
Save for the son that she did litter here,
A freckled whelp, hag-born—not honoured with
A human shape.
ARIEL   Yes, Caliban, her son.
PROSPERO
Dull thing, I say so: he, that Caliban

282 she] ROWE; he F

266 **for one thing she did** She was pregnant
(see l. 269), and pregnancy required the
commutation of a capital sentence.
Editorial debate over the *one thing* was
energetic until the early years of this century, most critics resisting the idea that
Sycorax's pregnancy saved her life. The
problematic element in the passage is not
its meaning, but the obliqueness of Prospero's reference to it.
269 **blue-eyed** generally explained as 'with
blue eyelids', implying pregnancy. W. A.
Wright compared the description of the
pregnant Duchess of Malfi, 'The fins of her
eyelids look most teeming blue' (2.1.67).
OED's citations for *blue-eyed* refer either to
the pupil or the area around the eye, but
the entry for *eye* notes that the word was
taken to include the eyelid as well. C. J.
Sisson argues unconvincingly in favour of
'tearstained, livid' (*New Readings*, i. 46).

273 **earthy** and therefore antithetical to
Ariel's volatile nature
274 **hests** behests
275 **ministers** agents
279 Sycorax, then, died sometime before
Prospero came to the island, and thus
more than twelve years ago. Caliban is
therefore at least twenty-four at the time
of the play, and was at least thirteen when
Prospero arrived with the three-year old
Miranda.
281 **as mill-wheels strike** as the blades of
water-wheels strike the water. Kittredge
suggested that the reference was to the
mill's clapper, a device which shakes the
hopper so as to move the grain down to
the millstones; but there is no evidence
that a clapper was ever referred to as a
mill-wheel.
285 **Dull thing, I say so** Prospero's vexation
continues: 'Don't parrot what I say!'

116

Whom now I keep in service. Thou best know'st
What torment I did find thee in. Thy groans
Did make wolves howl, and penetrate the breasts
Of ever-angry bears—it was a torment
To lay upon the damned, which Sycorax   290
Could not again undo. It was mine art,
When I arrived and heard thee, that made gape
The pine and let thee out.

ARIEL   I thank thee, master.

PROSPERO

If thou more murmur'st, I will rend an oak
And peg thee in his knotty entrails till
Thou hast howled away twelve winters.

ARIEL   Pardon, master.
I will be correspondent to command
And do my spriting gently.

PROSPERO   Do so, and after two days
I will discharge thee.

ARIEL   That's my noble master.
What shall I do? Say what: what shall I do?   300

PROSPERO

Go, make thyself like a nymph o'th'sea.

288–9 **penetrate the breasts | Of** i.e. arouse
   sympathy in
291–3 **Could . . . out** Prospero thus demonstrates that his magic is more powerful
   than Sycorax's.
297 **correspondent** responsive
298 **gently** graciously
301–4 There are obvious problems with F's
   text here. Dropping the meaningless *thine
   and* would rectify line 302, but is no help
   with 301 or 304. The editor of F2 tried to
   regularize the metre of 301 by printing
   'Go make thyself like to a nymph o'th'
   sea'. But clearly more has happened to
   the text than the omission of *to* and the
   addition of *thine and*, and felicitous
   emendation will not restore the original
   sense. F's repetition of the speech heading
   for Prospero after Ariel's exit may indicate
   that some sort of reply for Ariel (e.g. 'I
   go, I go': compare 4.1.187) has been
   omitted; but a longer exchange may also
   have been revised and cut down. If something has been eliminated, it would seem

to be Prospero's instructions regarding
the other spirits who are to be invoked in
Ariel's song as a sea-nymph, 'Full fathom
five'. These must be at least audibly
present for 'sweet sprites' to 'bear the burden' of the song 'dispersedly', as they do
at lines 381 ff. There is no indication in
the text as we have it that they physically
appear, but if they were originally present
either as dancers or an onstage chorus, it
is they who would have been 'subject to
no eye' but Ariel's and Prospero's. The
revision may have been made to reduce
the personnel required to perform the
play. Judging from its awkwardness, it is
unlikely to have been authorial. See the
Introduction, p. 61.
301 **like a nymph o'th' sea** The disguise is, of
   course, logically pointless if Ariel is invisible to everyone except Prospero. But
   he is visible to the audience, and the costume is the appropriate one to adopt in
   singing to Ferdinand on the shore. Compare Feste's curate disguise in *Twelfth*

Be subject to no sight but thine and mine, invisible
To every eyeball else. Go, take this shape,
And hither come in't; go! Hence, with diligence!

                                    Exit Ariel

(*To Miranda*) Awake, dear heart, awake. Thou hast
    slept well.
    Awake.

MIRANDA The strangeness of your story put
    Heaviness in me.

PROSPERO   Shake it off. Come on;
    We'll visit Caliban, my slave, who never
    Yields us kind answer.

MIRANDA   'Tis a villain, sir,
    I do not love to look on.

PROSPERO   But as 'tis,   310
    We cannot miss him. He does make our fire,
    Fetch in our wood, and serves in offices
    That profit us. What ho, slave! Caliban!
    Thou earth, thou, speak!

CALIBAN (*within*)   There's wood enough within.

PROSPERO
    Come forth, I say; there's other business for thee.
    Come, thou tortoise, when?
        Enter Ariel like a water-nymph
    Fine apparition! My quaint Ariel,
    Hark in thine ear. (*whispers*)

ARIEL   My lord, it shall be done.   *Exit*

---

304 *Exit Ariel*] ROWE; Exit F   305 *To Miranda*] This edition; *not in* F   Awake] ROWE;
Pro⟨spero⟩. Awake F (*repeating the speech heading*)   318 *whispers*] This edition; *not in* F

Night, and Maria's comment on its point-
lessness, 4.2.64–5.

307 **Heaviness** drowsiness

311 **miss** do without

314 **earth** in contrast to Prospero's other
    servant, the spirit of air
    within i.e. within the discovery place at
    the back of the stage. Caliban's 'hard

rock' dwelling (l. 343) may have been a
small movable property erected either in-
side or immediately in front of this alcove,
but the discovery place itself may just as
well have served as his den.

317 **quaint** The word includes the senses of
    ingenious and skilful, curious in ap-
    pearance, and elegant.

PROSPERO

Thou poisonous slave, got by the devil himself
Upon thy wicked dam, come forth!   320
  Enter Caliban

CALIBAN

As wicked dew as e'er my mother brushed
With raven's feather from unwholesome fen
Drop on you both! A south-west blow on ye
And blister you all o'er!

PROSPERO

For this be sure tonight thou shalt have cramps,
Side-stitches that shall pen thy breath up. Urchins
Shall, for that vast of night that they may work,
All exercise on thee. Thou shalt be pinched
As thick as honeycomb, each pinch more stinging
Than bees that made 'em.

CALIBAN   I must eat my dinner.   330
This island's mine by Sycorax my mother,
Which thou tak'st from me. When thou cam'st first,
Thou strok'st me and made much of me; wouldst give
  me
Water with berries in't, and teach me how
To name the bigger light and how the less,

---

327 Shall, . . . night . . . work,] ROWE 1714; Shall . . . night, . . . worke F

319–20 **got by the devil himself | Upon thy wicked dam** alluding to stories of sexual liaisons between witches and the devil. The identity of Caliban's father is mentioned nowhere else in the play, though Prospero calls him a 'demi-devil' and a bastard at 5.1.272–3. It is not clear whether Prospero's expostulation is mere invective or a literal account of Caliban's conception; and indeed, its dramatic purpose is amply served by leaving us in doubt. See the Introduction, p. 25.

321 **wicked** both harmful and foul

321–2 **dew . . . raven's feather** Dew was a common ingredient of magical potions, required by Prospero as well as Sycorax: see l. 228. The raven was especially associated with witchcraft, and its Greek and Latin name, *korax/corax*, is clearly related to the unexplained name Sycorax.

323 **south-west** Southerly winds were associated with warm, damp weather,

and considered unwholesome.

326 **Urchins** goblins, so called 'from the supposition that they occasionally assumed the form of a hedgehog' (*OED*). Compare 2.2.5.

327 **vast** great stretch (normally of space, not time)

328–9 **pinched | As thick as honeycomb** covered with pinches as thoroughly as the honeycomb has cells; the image perhaps derives from the notion that bees mould their wax by pinching it into shape.

331 Caliban bases his claim to the island on inheritance. If he is, as Prospero asserts, illegitimate, the claim would be invalid. See the Introduction, p. 25.

333 **strok'st** for 'strok'dst': Abbott 473

335 **bigger . . . less** echoing Genesis 1 : 16; in the Geneva Bible, 'God then made two great lights: the greater light to rule the day, and the less light to rule the night'. No English Bible reads 'bigger'.

That burn by day and night; and then I loved thee,
And showed thee all the qualities o'th' isle,
The fresh springs, brine pits, barren place and fertile—
Cursed be I that did so! All the charms
Of Sycorax, toads, beetles, bats light on you!   340
For I am all the subjects that you have,
Which first was mine own king, and here you sty me
In this hard rock, whiles you do keep from me
The rest o'th' island.

PROSPERO   Thou most lying slave,
    Whom stripes may move, not kindness, I have used
        thee—
Filth as thou art— with humane care, and lodged thee
In mine own cell, till thou didst seek to violate
The honour of my child.

CALIBAN   O ho, O ho! Would't had been
        done!
● Thou didst prevent me—I had peopled else
This isle with Calibans.

MIRANDA   Abhorrèd slave,   350
    Which any print of goodness wilt not take,
Being capable of all ill! I pitied thee,
Took pains to make thee speak, taught thee each hour
One thing or other. When thou didst not, savage,
Know thine own meaning, but wouldst gabble like
A thing most brutish, I endowed thy purposes
With words that made them known. But thy vile
        race—
Though thou didst learn—had that in't which good
        natures

---

339 Cursed . . . did] F1 ; Curs'd be I that I did F2

339 **charms** spells
342 **sty me** pen me up like a pig
346 **humane** F's spelling, not distinguished
    in the period from *human*, and accented
    until the eighteenth century on the first
    syllable
350–61 **Abhorrèd slave . . . prison** From
    Dryden to Kittredge, this speech was al-
    most always reassigned to Prospero, its
    tone being considered inappropriate to
    Miranda's character. But the energy
    displayed here may be taken as an impor-
tant aspect of her nature, in evidence
again when she defends Ferdinand from
her father's charges at the end of Act 1.
See the Introduction, p. 17.
351 **print** imprint. The metaphor alludes at
    once to coinage, wax seals, and
    typography.
352 **capable of** susceptible (only) to
357 **race** 'Natural or inherited disposition'
    (*OED*). Compare *Measure* 2.4.160: 'Now
    I give my sensual race the rein.'

Could not abide to be with; therefore wast thou
Deservedly confined into this rock,   360
Who hadst deserved more than a prison.

CALIBAN

You taught me language, and my profit on't
Is I know how to curse. The red plague rid you
For learning me your language!

PROSPERO   Hag-seed, hence!
Fetch us in fuel, and be quick, thou'rt best,
To answer other business—shrug'st thou, malice?
If thou neglect'st, or dost unwillingly
What I command, I'll rack thee with old cramps,
Fill all thy bones with achës, make thee roar,
That beasts shall tremble at thy din.

CALIBAN   No, pray thee.   370
(*Aside*) I must obey. His art is of such power,
It would control my dam's god Setebos
And make a vassal of him.

PROSPERO   So, slave, hence!

Exit Caliban

*Enter Ferdinand, and Ariel invisible, playing and
singing*

ARIEL (*sings*)

Come unto these yellow sands,
And then take hands;

374, 397 ARIEL (*sings*)] This edition; *Ariel* Song F

---

363 **red** 'Applied to various diseases marked
by evacuation of blood or cutaneous erup
tions' (*OED* 16b). Thersites invokes a red
murrain on Ajax, *Troilus* 2.1.19.
  rid kill, destroy (*OED* IIc)
364 **learning** teaching
366 **answer other business** perform other
tasks
368 **old cramps** the cramps of old age
369 **achës** Until the seventeenth century,
the noun was pronounced *atch*, the verb
ake (*OED*).
372 **Setebos** The name is found in accounts
of Magellan's voyages as that of a 'great
devil' of the Patagonians (e.g. Richard
Eden, *The Decades of the New World of West*

India (London, 1555), p. 219ᵛ). See
Charles Frey, '*The Tempest* and the New
World', *ShQ* 30, 1 (1979), p. 29.
373.2 **invisible** Ariel is dressed as a sea-
nymph; his invisibility is presumably
established simply by Prospero's instructions at l. 302 and by Ferdinand's failure
to see him. Henslowe's papers list as a
property of the Lord Admiral's Men a
'robe for to goo invisibell' (eds. Foakes
and Rickert, p. 325), but such a garment
would be unnecessary here, and awk-
ward in conjunction with the nymph's
wings.
  playing probably a lute; later Ariel plays
a tabor and pipe (3.2.122.1).

  Curtsied when you have, and kissed
   The wild waves whist,
  Foot it featly here and there,
  And sweet sprites bear
  The burden. Hark, hark!   380
      (*Burden, dispersedly*) Bow-wow.
   The watch dogs bark.
      (*Burden, dispersedly*) Bow-wow.
  Hark, hark! I hear
  The strain of strutting Chanticleer
  Cry cock a diddle dow.
     (*Burden, dispersedly*) Cock a diddle dow.

FERDINAND

  Where should this music be?—i'th' air or th' earth?
  It sounds no more; and sure it waits upon
  Some god o'th' island. Sitting on a bank,   390
  Weeping again the King my father's wreck,
  This music crept by me upon the waters,
  Allaying both their fury and my passion
  With its sweet air. Thence I have followed it,
  Or it hath drawn me rather; but 'tis gone.
  No, it begins again.

381 *Burden, dispersedly*] CAPELL; *before* Hark, hark! *in* F 383 *Burden, dispersedly*] CAPELL; *not in,* F 387 *Burden . . . dow*] This edition; *not in* F

376–7 kissed . . . whist either 'kissed the wild waves into silence' or 'kissed (each other) until the wild waves are silent'. The usual editorial insertion of a comma after *kissed* is entirely unwarranted, and weakens the sense. Alan Brissenden observes that the emendation 'is largely based on the mistaken belief that kissing was a customary part of the beginning of dances'. He finds 'no evidence that kissing was a usual part of the honour ("curtsy", "bow", "reverence") which preceded a dance, or that a kiss normally came at the end' (*Shakespeare and the Dance*, Atlantic Highlands, N. J., 1981), p. 97).
378 featly both neatly and elegantly

381 burden refrain
  dispersedly i.e. not in unison
386 F's indication of the burdens is confusing and, as it stands, must be inaccurate. The song must have had a final refrain, but whether *Cry* is a stage direction, or, if not, whether anything more than *Cry* in this line belongs to Ariel, is impossible to determine.
389 waits attends
390–1 Sitting . . . Weeping technically dangling modifiers. One would expect a clause beginning with 'I' after them, and the switch of subject emphasizes Ferdinand's passivity.
393 passion literally, suffering

ARIEL (*sings*)

> Full fathom five thy father lies,
> Of his bones are coral made;
> Those are pearls that were his eyes;
> Nothing of him that doth fade,   400
> But doth suffer a sea-change
> Into something rich and strange.
> Sea-nymphs hourly ring his knell.
>   (*Burden*) Ding dong.
> Hark, now I hear them, ding dong bell.

FERDINAND

The ditty does remember my drowned father.
This is no mortal business, nor no sound
That the earth owes—I hear it now above me.

PROSPERO (*to Miranda*)

The fringéd curtains of thine eye advance,
And say what thou seest yond.

MIRANDA   What is't?—a spirit?   410

Lord, how it looks about! Believe me, sir,
It carries a brave form. But 'tis a spirit.

PROSPERO

No, wench, it eats and sleeps, and hath such senses
As we have—such. This gallant which thou seest

---

397 ff. What is almost certainly the original music for Ariel's song, by Robert Johnson, is in Appendix C.

397 **fathom** originally the measure of a man's outstretched arms from fingertip to fingertip, reckoned as 6 feet. The drowned father is thus 30 feet deep.

398 **are** The verb takes its number from the apparent subject preceding it. See G. L. Brook, *The Language of Shakespeare*, 51 (d).

404 (**Burden**) **Ding dong** F sets this off from the text of the song, and prints it in Roman type—the song itself is in italic. Possibly it is not part of the song, but a marginal direction for a sound effect.

406 **ditty** a song, especially its verbal element

remember commemorate

407 **mortal** both human, and pertaining to death: Ferdinand's perception grows as he muses on the song.

408 **owes** owns

I hear it now above me Either Ariel moves to an upper portion of the stage and continues playing, or the music now comes from a consort of musicians in the music house above.

409 Presumably Miranda has averted her eyes from the action after rebuking Caliban. For the locution, compare *Pericles* 3.2.99 ff.: 'Her eyelids, cases to those heavenly jewels | . . . Begin to part their fringes of bright gold.' *Advance* = raise (*OED* iii. 9); the *curtains* suggest the theatrical metaphor of a discovery scene, but stage curtains in the period were either drawn aside on traverses or dropped from above. Inigo Jones did not devise a curtain that could be raised until the 1630s.

413 **wench** See note on 1.2.139.

414 **gallant** 'A man of fashion and pleasure; a fine gentleman', especially 'a ladies' man' (*OED* B1, 3), often, as here, with playful or semi-ironic overtones

Was in the wreck, and but he's something stained
With grief—that's beauty's canker—thou mightst call
   him
A goodly person. He hath lost his fellows,
And strays about to find 'em.

MIRANDA   I might call him
A thing divine, for nothing natural
I ever saw so noble.

PROSPERO *(aside)*   It goes on, I see,   420
As my soul prompts it. *(To Ariel)* Spirit, fine spirit, I'll
   free thee
Within two days for this.

FERDINAND   Most sure, the goddess
On whom these airs attend. Vouchsafe my prayer
May know if you remain upon this island,
And that you will some good instruction give
How I may bear me here. My prime request,
Which I do last pronounce, is—O you wonder!—
If you be maid or no?

MIRANDA   No wonder, sir,
But certainly a maid.

FERDINAND   My language! Heavens!
I am the best of them that speak this speech,   430
Were I but where 'tis spoken.

PROSPERO   How? The best?

---

421 *To Ariel*] This edition; *not in* F

415 **but** except for the fact that
  something somewhat
416 **grief—that's beauty's canker** A canker
  is variously a spreading sore; rust; a
  disease, especially of fruit trees; a destruc-
  tive larva; and, from these, the general
  sense of 'anything that frets, corrodes,
  corrupts or consumes slowly and secretly'
  (*OED* 6); *grief—that's beauty's canker*
  means either that grief is a disease especially disfiguring to the beautiful, or that
  grief is attracted especially to the beautiful
  as the larva is to the rose, precisely
  because of its beauty.
419 **natural** as opposed to the artificial crea-
  tions of, e.g., Prospero's masque. Kermode glosses 'in the realm of nature, as
  opposed to the realm of spirit', but the
  spirits of the play *are* spirits of nature.

420 **It** my plan
422 **Most sure, the goddess** translating
  Aeneas' reaction to the sight of Venus
  after the Trojan shipwreck, 'O dea certe'
  (*Aeneid*, i. 328). On the undercurrent of
  allusions to the *Aeneid* throughout the
  play, see the Introduction, pp. 39–42.
423 **airs** Ariel's songs (compare 392–4)
424 **May know** that I may know; for the
  ellipsis, see Abbott, p. 382.
  remain dwell (*OED* 4b)
426 **bear me** conduct myself
427 **wonder** The epithet puns on Miranda's
  name.
428 **maid** a girl, as opposed to either a goddess or a married woman
430 Ferdinand assumes that he has succeeded to his father's throne.

What wert thou if the King of Naples heard thee?

FERDINAND

A single thing, as I am now, that wonders
To hear thee speak of Naples. He does hear me,
And that he does, I weep: myself am Naples,
Who with mine eyes, never since at ebb, beheld
The King my father wrecked.

MIRANDA   Alack, for mercy!

FERDINAND

Yes, faith, and all his lords, the Duke of Milan
And his brave son being twain.

PROSPERO (*aside*)   The Duke of Milan
And his more braver daughter could control thee   440
If now 'twere fit to do't. At the first sight
They have changed eyes. Delicate Ariel,
I'll set thee free for this.—A word, good sir:
I fear you have done yourself some wrong; a word.

MIRANDA

Why speaks my father so ungently? This
Is the third man that e'er I saw, the first
That e'er I sighed for. Pity move my father
To be inclined my way!

FERDINAND   O, if a virgin,

---

433 **A single thing, as I am now** The King of
Naples and I are one and the same; I
would be only what I am now. *Single* has
overtones of simplicity and poverty, but
the word also, in conjunction with
Ferdinand's question about whether
Miranda is a maid, asserts that he too is
unmarried.

434, 435 **Naples** the King of Naples

434 **He does hear me** i.e. because I am he

436 **never since at ebb** continually weeping

439 **his brave son** Antonio's son is mentioned nowhere else in the play, and
cannot be identified with any of the shipwreck victims. Dover Wilson argued that
he is a vestige of an earlier version of the
play, Kermode that 'Sh. began writing
with a somewhat hazy understanding of
the dynastic relationships he was to deal
with'. Possibly a parallel to Ferdinand
was originally contemplated by Shakespeare, and then abandoned as the drama
took shape. Attempts to explain the

allusion by making *his* refer to *the King my
father* in l. 437 are obviously misguided:
Ferdinand would not characterize himself
as *brave*.

440 **more braver** Abbott 11
control challenge, refute (*OED* 3b)

441–2 **At the first sight . . . eyes** Compare
Marlowe's *Hero and Leander* i. 175 : 'Who-
ever loved that loved not at first sight?',
quoted in *As You Like It* 3.5.81 and
elsewhere. Tilley's citation, 'Love not at
the first look' (L426), cautions against
such passions, but Dent doubts that this
should be considered an authentic
proverb.

442 **changed** interchanged. 'they can't take
their eyes off one another.'
Delicate graceful, artful

444 **you have done yourself some wrong** i.e.
in claiming to be King of Naples. The tone
is ironic, but Prospero is, of course, quite
correct.

448 **if a virgin** Ferdinand ignores Prospero

And your affection not gone forth, I'll make you
The Queen of Naples.

PROSPERO   Soft, sir, one word more.   450
  (*Aside*) They are both in either's powers; but this swift
    business
  I must uneasy make lest too light winning
  Make the prize light.—One word more: I charge thee
  That thou attend me. Thou dost here usurp
  The name thou ow'st not, and hast put thyself
  Upon this island as a spy, to win it
  From me, the lord on't.

FERDINAND   No, as I am a man!

MIRANDA
  There's nothing ill can dwell in such a temple.
  If the ill spirit have so fair a house,
  Good things will strive to dwell with't.

PROSPERO   Follow me.—   460
  Speak not you for him: he's a traitor.—Come,
  I'll manacle thy neck and feet together.
  Sea-water shalt thou drink; thy food shall be
  The fresh-brook mussels, withered roots, and husks
  Wherein the acorn cradled. Follow.

FERDINAND   No;
  I will resist such entertainment till
  Mine enemy has more power.
                    He draws, and is charmed from moving

MIRANDA   O dear father,

and returns to his 'prime request' of
ll. 426–8. Ellipses after *if* were common:
Abbott 387.

452 **uneasy** difficult

452–3 **light . . . light** easy . . . cheap, with
perhaps an overtone of 'promiscuous' in
the second instance. Prospero's explanation of his behaviour has generally been
found unconvincing.

454–5 **usurp | The name thou ow'st not**
Again, the charge is literally true, though
what follows from it is Prospero's invention. See the Introduction, p. 29.

455 **ow'st** own'st

458 **temple** 'Any place regarded as occupied
by the divine presence; *spec.* the person or

body of a Christian' (*OED* 3)

459–60 **If the ill spirit . . . dwell with't** and,
being stronger, expel it. Miranda's assertion is conventional Renaissance Platonic
doctrine; equally conventional in the
period, however, are observations about
the deceptiveness of attractive exteriors.
In the dramatic circumstances, Miranda's
speech expresses more naïveté than
Platonism. Compare 5.1.181–4.

464 **fresh-brook mussels** Fresh-water mussels are inedible. *OED* cites Owen's *Pem-
brokeshire*, 1603: 'The Ryver muskles are
not for meate.'

466 **entertainment** treatment (*OED* 5); Ferdinand's usage is not necessarily ironic.

Make not too rash a trial of him, for
He's gentle, and not fearful.
PROSPERO   What, I say—
My foot my tutor? Put thy sword up, traitor,   470
Who mak'st a show but dar'st not strike, thy conscience
Is so possessed with guilt. Come from thy ward,
For I can here disarm thee with this stick
And make thy weapon drop.
MIRANDA   Beseech you, father—
PROSPERO
Hence! Hang not on my garments.
MIRANDA   Sir, have pity;
I'll be his surety.
PROSPERO   Silence! One word more
Shall make me chide thee, if not hate thee. What,
An advocate for an impostor? Hush!
Thou think'st there is no more such shapes as he,
Having seen but him and Caliban. Foolish wench,   480
To th' most of men this is a Caliban,
And they to him are angels.
MIRANDA   My affections
Are then most humble. I have no ambition
To see a goodlier man.
PROSPERO (*to Ferdinand*) Come on, obey.
Thy nerves are in their infancy again
And have no vigour in them.
FERDINAND   So they are.
My spirits, as in a dream, are all bound up.
My father's loss, the weakness which I feel,
The wreck of all my friends, nor this man's threats,
To whom I am subdued, are but light to me,   490
Might I but through my prison once a day

---

469 **gentle, and not fearful** noble, and
    therefore not a coward
470 **My foot my tutor?** 'Shall the lowest of
    my appendages teach me how to act?'
    Proverbial: Tilley cites 'Do not make the
    foot the head' (F562).
472 **ward** defensive posture
479 **there is no more such shapes** Abbott
    335
481 **To** compared to

485 **nerves** sinews
488–90 **My father's loss . . . light** There is a
    good deal of confusion in the syntax, as
    there is in Ferdinand's mind: *nor* (489)
    introduces a double construction: (a) my
    father's loss, the weakness, etc., are light;
    (b) neither my father's loss . . . nor this
    man's threats . . . are anything but light.
    (Compare Abbott 408.)
490 **but** merely, otherwise than

Behold this maid. All corners else o' th' earth
Let liberty make use of—space enough
Have I in such a prison.

PROSPERO (*aside*)   It works. (*To Ferdinand*) Come on.—
(*To Ariel*) Thou hast done well, fine Ariel. Follow me;
Hark what thou else shalt do me.

MIRANDA (*to Ferdinand*)   Be of comfort.
My father's of a better nature, sir,
Than he appears by speech. This is unwonted
Which now came from him.

PROSPERO (*to Ariel*)   Thou shalt be as free
As mountain winds; but then exactly do   500
All points of my command.

ARIEL   To th' syllable.

PROSPERO (*to Ferdinand*)
Come, follow. (*To Miranda*)—Speak not for him. *Exeunt*

2.1   *Enter Alonso, Sebastian, Antonio, Gonzalo, Adrian,*
       Francisco

GONZALO (*to Alonso*)
Beseech you, sir, be merry. You have cause—
So have we all—of joy, for our escape
Is much beyond our loss. Our hint of woe
Is common: every day some sailor's wife,
The masters of some merchant, and the merchant
Have just our theme of woe; but for the miracle—

---

494 *To Ferdinand*] This edition; *not in* F
   2.1.0.1–2 *Enter . . . Francisco*] *see note*

492 **All corners** any parts 'whatsoever, even
     the smallest, most distant and secluded'
     (*OED* 7)
495–6 **Follow . . . do me** The Cambridge
     editors had the whole sentence addressed
     to Ferdinand; the Riverside addresses
     Follow me to Ferdinand and the rest to
     Ariel. Both are attractive, if arbitrary,
     editorial decisions.
496 **do me** do for me
499–500 **as free | As mountain winds** in
     Ariel's case, a strikingly literal realization
     of the proverbial 'as free as the air' (Tilley
     A88)
500 **then** i.e. if that is to be so
2.1.0.1–2 Here and at 3.3.0.1 the opening

stage directions add *and others* and *etc.*;
but when the courtiers reassemble at
5.1.57.2–4, the stage direction makes
no such stipulation. Both indications
cannot be correct, and I have assumed
that the supernumerary characters were
erroneously included in Crane's text.
3 **beyond** more important than
   hint occasion
5 **The masters . . . the merchant** either the
   officers or the owners of some merchant-
   man, and either the vessel itself or the
   merchant to whom the cargo belongs.
   The line has produced a great deal of
   editorial quibbling, for the most part over
   the fact that, if the second interpretation

I mean our preservation—few in millions
Can speak like us. Then wisely, good sir, weigh
Our sorrow with our comfort.

ALONSO   Prithee, peace.

SEBASTIAN *(aside to Antonio)* He receives comfort like cold   10
porridge.

ANTONIO The visitor will not give him o'er so.

SEBASTIAN Look, he's winding up the watch of his wit. By
and by it will strike.

GONZALO Sir,—

SEBASTIAN One. Tell.

GONZALO   —when every grief is entertained
That's offered, comes to th' entertainer—

SEBASTIAN A dollar.

GONZALO Dolour comes to him indeed. You have spoken   20
truer than you purposed.

SEBASTIAN You have taken it wiselier than I meant you
should.

GONZALO Therefore, my lord,—

ANTONIO Fie, what a spendthrift is he of his tongue!

ALONSO *(to Gonzalo)* I prithee, spare.

GONZALO Well, I have done. But yet—

SEBASTIAN He will be talking.

ANTONIO Which, of he or Adrian, for a good wager, first
begins to crow?   30

SEBASTIAN The old cock.

of *merchant* is correct, Shakespeare was using the word in two different senses in the same line, 'not as a pun, but from carelessness' (Kermode). This conclusion is not inevitable, but if it were, it would not be an argument in favour of any of the emendations that have been proposed.

9, 10 **peace, porridge** punning on 'pease-porridge'

10–12 Antonio's and Sebastian's opening exchange should probably be treated as a private conversation; but for the most part throughout this scene their insulting banter is designed to be overheard.

11 **visitor** Gonzalo is compared with the church functionary charged with comforting the sick of the parish.
give him o'er leave him alone

12–13 **watch... strike** 'From the beginning

of the seventeenth century "watches" (from the context clearly pocket watches) are often spoken of as striking' (*OED* s. watch iv. 21).

16 **One. Tell.** His watch has struck one. Keep count.

17–18 **entertained | That's offered** Most editors end the line at *offered*, but F's lineation needs no emendation. The blank verse line is 'Sir, when every grief is entertained'; it has been interrupted by Sebastian's 'One. Tell.' Gonzalo attempts to ignore the interruption and continues.

19 **A dollar** i.e. in payment: Sebastian quibbles on *entertainer* = performer.

25 **fie... tongue** proverbial (Dent T64.1.)

29–32 **which, ... cockerel** Compare the proverb 'the young cock crows as he the old hears' (Tilley C491).

ANTONIO The cockerel.

SEBASTIAN Done. The wager?

ANTONIO A laughter.

SEBASTIAN A match!

ADRIAN Though this island seem to be desert—

ANTONIO Ha, ha, ha!

SEBASTIAN So, you're paid!

ADRIAN Uninhabitable, and almost inaccessible—

SEBASTIAN Yet—   40

ADRIAN Yet—

ANTONIO He could not miss't.

ADRIAN It must needs be of subtle, tender, and delicate temperance.

ANTONIO Temperance was a delicate wench.

SEBASTIAN Ay, and a subtle, as he most learnedly delivered.

ADRIAN The air breathes upon us here most sweetly.

SEBASTIAN As if it had lungs, and rotten ones.

ANTONIO Or as 'twere perfumed by a fen.

GONZALO Here is everything advantageous to life.   50

ANTONIO True, save means to live.

SEBASTIAN Of that there's none or little.

GONZALO How lush and lusty the grass looks! How green!

ANTONIO The ground indeed is tawny.

SEBASTIAN With an eye of green in't.

---

37 ANTONIO] GRANT WHITE; Seb⟨astian⟩. F   38 SEBASTIAN] GRANT WHITE; Ant⟨onio⟩. F

34 **A laughter** The proverb is 'he laughs that wins' (Tilley L93); a laughter is also 'the whole number of eggs laid by a fowl before she is ready to sit' (*OED* s. laughter 2).

36 **desert** uninhabited

37–8 **Ha . . . paid** F gives the laugh to Sebastian, l. 38 to Antonio. But, since Antonio has won the bet and the prize is a laugh, it seems clear that the speech headings have been reversed. Sebastian's line means 'you've had your laugh'.

43 **subtle** gentle

44 **temperance** 'mildness of weather or climate' (*OED* 4)

45 **Temperance . . . wench** Antonio takes Temperance to be a girl's name; hence *delicate* = given to pleasure.

46 **subtle** Sebastian develops the theme of delicacy: *subtle* here implies craftiness and (sexual) expertise; and, with *learned-*ly, plays on the sense of acute or speculative.

53 **lush** The relevant meaning current in Shakespeare's time is 'soft, tender'. *OED* records the sense 'succulent and luxuriant in growth' (erroneously) in this passage, and otherwise only in nineteenth-century examples from Keats onwards which evidently derive from it. Shakespeare does not use the word elsewhere.

54 **The . . . tawny** not a contradiction: Antonio is mocking Gonzalo's compulsion to remark on everything.

55 **eye** 'slight shade, tinge' (*OED* ii. 9). This passage is the earliest *OED* citation, but the entry gives other seventeenth-century examples from Suckling, Fuller, and Evelyn. Compare French *œil*, common in this sense.

ANTONIO He misses not much.

SEBASTIAN No, he doth but mistake the truth totally.

GONZALO But the rarity of it is, which is indeed almost beyond credit—

SEBASTIAN As many vouched rarities are. 60

GONZALO That our garments, being, as they were, drenched in the sea, hold notwithstanding their freshness and gloss, being rather new-dyed than stained with salt water.

ANTONIO If but one of his pockets could speak, would it not say he lies?

SEBASTIAN Ay, or very falsely pocket up his report.

GONZALO Methinks our garments are now as fresh as when we put them on first in Afric, at the marriage of the King's fair daughter Claribel to the King of Tunis. 70

SEBASTIAN 'Twas a sweet marriage, and we prosper well in our return.

ADRIAN Tunis was never graced before with such a paragon to their queen.

GONZALO Not since widow Dido's time.

ANTONIO Widow? A pox o' that. How came that widow in? Widow Dido!

SEBASTIAN What if he had said 'widower Aeneas' too? Good lord, how you take it!

ADRIAN 'Widow Dido' said you? You make me study of 80 that. She was of Carthage, not of Tunis.

---

63 gloss] DYCE 2; glosses F

---

56 **He . . . much** i.e. Gonzalo's is the 'eye of green'.

58, 60 **rarity, rarities** exceptional quality, unique phenomena

61–4 **our garments . . . water** Compare Ariel, 1.2.218–19: 'On their sustaining garments not a blemish, | But fresher than before'.

63 **gloss** I assume F's 'glosses' is a misreading' of 'glosse'; in Crane's handwriting it is an easy one. Shakespeare nowhere else uses 'gloss' in the plural.

65–6 **If . . . lies** Since Ariel has testified to the condition of the garments, Antonio is presumably being merely perverse; but the line also contributes to a general sense that the quality of the island and of experience on it is perceived diversely and subjectively by the various characters.

67 **pocket up** conceal or suppress

his report either Gonzalo's reputation, or the report of it. For *his* = its, see Abbott 228.

77, 78 **Widow Dido, widower Aeneas** Dido was the widow of Sychaeus; Aeneas' wife Creusa died in the sack of Troy. Recollections of and allusions to the *Aeneid* provide an important undercurrent throughout the play. See the Introduction, pp. 40–2.

80 **study of** meditate on (*OED* 2)

GONZALO This Tunis, sir, was Carthage.

ADRIAN Carthage?

GONZALO I assure you, Carthage.

ANTONIO His word is more than the miraculous harp.

SEBASTIAN He hath raised the wall, and houses too.

ANTONIO What impossible matter will he make easy next?

SEBASTIAN I think he will carry this island home in his
pocket and give it his son for an apple.

ANTONIO And sowing the kernels of it in the sea, bring forth   90
more islands.

GONZALO Ay.

ANTONIO Why, in good time.

GONZALO (*to Alonso*) Sir, we were talking that our garments
seem now as fresh as when we were at Tunis at the
marriage of your daughter, who is now queen.

ANTONIO And the rarest that e'er came there.

SEBASTIAN Bate, I beseech you, widow Dido.

ANTONIO O, widow Dido? Ay, widow Dido.

GONZALO Is not, sir, my doublet as fresh as the first day I   100
wore it? I mean, in a sort.

ANTONIO That sort was well fished for.

GONZALO When I wore it at your daughter's marriage.

ALONSO
You cram these words into mine ears against
The stomach of my sense. Would I had never

82 **This . . . was Carthage** Gonzalo is correct
in the sense that, though Carthage and
Tunis were always separate cities, after
the destruction of Carthage Tunis took its
place as the political and commercial
centre of the region. This is presumably
what Antonio and Sebastian are
quibbling about. See the Introduction,
p. 42, n. 2.

85–6 **His . . . too** Amphion's music raised
only the walls of Thebes, but Gonzalo has
rebuilt the whole city of Carthage.

88–90 **I think . . . sea** Compare *Antony and
Cleopatra* 5.20.90–1: 'Realms and is-
lands were | As plates dropped from his
pockets.'

92 **Ay** Since Rowe, though with the notable
exception of Johnson, F's 'I.' has been
regularly interpreted as Gonzalo's af-
firmation of his last assertion, l. 82, and I
have followed the critical consensus. But

Johnson's version is attractive: he moder-
nized the line to 'I—', and assumed that
Gonzalo is beginning a new sentence
which is at once rudely interrupted.
Michael J. Warren calls attention to the
difficulties of modernizing the line in
'Textual Problems, Editorial Assertions in
Editions of Shakespeare', in *Textual Criti-
cism and Literary Interpretation*, ed. Jerome
J. McGann (Chicago, 1985), p. 33.

93 **in good time** ironic: 'at long last'.

98 **Bate . . . Dido** either 'except widow Dido'
or 'don't mention widow Dido again'

101 **in a sort** to some extent, in some way
(*OED* iii. 21c, d)

102 **sort** lot (*OED* sb¹, 2): Antonio's
metaphor is of drawing lots.

104–5 **You . . . sense** Alonso complains that
he is being force-fed; *stomach* = temper,
disposition (*OED* 7); *sense* means both
intention and perception.

Married my daughter there, for coming thence
My son is lost, and, in my rate, she too,
Who is so far from Italy removed
I ne'er again shall see her. O thou mine heir
Of Naples and of Milan, what strange fish   110
Hath made his meal on thee?
FRANCISCO   Sir, he may live.
I saw him beat the surges under him
And ride upon their backs; he trod the water,
Whose enmity he flung aside, and breasted
The surge most swoll'n that met him; his bold head
'Bove the contentious waves he kept, and oared
Himself with his good arms in lusty stroke
To th' shore, that o'er his wave-worn basis bowed,
As stooping to relieve him. I not doubt
He came alive to land.
ALONSO   No, no, he's gone.   120
SEBASTIAN
Sir, you may thank yourself for this great loss,
That would not bless our Europe with your daughter,
But rather lose her to an African,
Where she, at least, is banished from your eye,
Who hath cause to wet the grief on't.
ALONSO   Prithee, peace.
SEBASTIAN
You were kneeled to and importuned otherwise
By all of us, and the fair soul herself

---

107 **rate** estimation (*OED* 2b)

111–20 **Sir . . . land** The contrast of tone is important. Alonso's crudeness is answered with the high, heroic rhetoric of the Roman plays. Compare the account of Caesar and Cassius swimming the Tiber: 'The torrent roared, and we did buffet it | With lusty sinews, throwing it aside | And stemming it with hearts of controversy' (*Julius Caesar* 1.2.107–9).

116 **oared** *OED* records only this and a 1647 use of *oar* as a verb before the eighteenth century.

118 **his** its
basis the foot of the cliff above the shore. 'The image is of cliffs eroded by the surf at

their base, and so seeming to bend over the sea compassionately' (Barton).

119 **I not doubt** Abbott, 305

123 **lose** F's 'loose' is the normal spelling of *lose*, but carries the additional sense of 'release' or 'send forth': compare Polonius' plans for Ophelia, *Hamlet* 2.2.162: 'At such a time I'll lose my daughter to him'. The latter usage, if it is intended, would be ironic here, implying setting an imprisoned creature free.

125 **wet the grief on't** weep over the sorrow of it

126 **importuned** accented on the second syllable

Weighed between loathness and obedience at
Which end o'th' beam should bow. We have lost your
    son,
I fear, for ever. Milan and Naples have   130
More widows in them of this business' making
Than we bring men to comfort them.
The fault's your own.

ALONSO   So is the dear'st o'th' loss.

GONZALO
    My lord Sebastian,
    The truth you speak doth lack some gentleness,
    And time to speak it in—you rub the sore
    When you should bring the plaster.

SEBASTIAN   Very well.

ANTONIO
    And most chirurgeonly!

GONZALO (*to Alonso*)
    It is foul weather in us all, good sir,
    When you are cloudy.

SEBASTIAN   Foul weather?

ANTONIO   Very foul.   140

GONZALO (*to Alonso*)
    Had I plantation of this isle, my lord,—

ANTONIO
    He'd sow't with nettle-seed.

SEBASTIAN   Or docks, or mallows.

128–9 **Weighed . . . bow** *Weigh* is used both
  in its literal sense and in the sense of
  'ponder, consider' (*OED* 14b). The construction is a double one, combining
  'hung balanced between loathing and
  obedience' and 'pondered at which end of
  the (scale's) beam (she) should bow'.
132 **Than we bring men** Sebastian assumes
  both their return from the island and the
  loss of the rest of the fleet.
133 **dear'st** *dear* = 'precious in import or
  significance' (*OED* a¹, ii. 4b), costly (ii. 6),
  and also 'hard, severe, heavy, grievous'
  (a², 2): the last is apparently a different
  word.
136 **time** appropriate time
  rub the sore proverbial: Tilley S649
137 **plaster** 'a healing or soothing means or
  measure' (*OED* i. 1b)

138 **chirurgeonly** like a surgeon
140 **Foul . . . foul** F spells 'Fowle . . . foule',
  suggesting that some pun is intended.
  Possibly the play is on 'fool': Kökeritz (pp.
  75, 247) gives several examples of a *fool–
  foul–fowl* pun, though Cercignani considers this dubious. Alternatively, since
  Gonzalo has been described as 'the old
  cock' (l. 31), perhaps Antonio now recalls
  that characterization by mimicking a
  fowl.
141 **plantation** the right to colonize (see *OED*
  1c); Antonio and Sebastian take it in the
  sense of 'planting'.
142 **docks** 'coarse weedy herbs' (*OED* 1),
  characterized as 'hateful' in *Henry V*
  5.2.52 because they are, as here, inimical
  to productive cultivation. But *dock* is also
  well-known as the popular antidote for

GONZALO
—And were the king on't, what would I do?

SEBASTIAN
'Scape being drunk, for want of wine.

GONZALO
I'th' commonwealth I would by contraries
Execute all things, for no kind of traffic
Would I admit; no name of magistrate;
Letters should not be known; riches, poverty,
And use of service, none; contract, succession,
Bourn, bound of land, tilth, vineyard, none;   150
No use of metal, corn, or wine, or oil;
No occupation, all men idle, all,
And women too, but innocent and pure;
No sovereignty—

SEBASTIAN   Yet he would be king on't.

ANTONIO The latter end of his commonwealth forgets the
beginning.

GONZALO

All things in common nature should produce
Without sweat or endeavour. Treason, felony,

---

148 riches, poverty] F; poverty, riches CAPELL

nettle-stings' (*OED* 1a), hence presumably the association of the two as complementary weeds.
  mallows another weed, but also another antidote for Gonzalo's nettles: mallow roots were used to make a soothing ointment (Grigson, *The Englishman's Flora* (London, 1955), p. 100).
145–66 This passage is closely related to a section of Montaigne's essay 'Of the Cannibals' in John Florio's translation (1603): 'It is a nation . . . that hath no kind of traffic, no knowledge of letters, no intelligence of numbers, no name of magistrate nor of politic superiority, no use of service, of riches or of poverty, no contracts, no successions, no dividences, no occupation but idle, no respect of kindred but common, no apparel but natural, no manuring of lands, no use of wine, corn, or metal. The very words that import lying, falsehood, treason, dissimulation, covetousness, envy, detraction, and pardon were never heard of amongst them.' (There is another passage

from Montaigne at 5.1.27–8.) See the Introduction, pp. 34–6, and Appendix D.
145 **by contraries** in a manner opposite to what is usual
146 **traffic** commerce
148 **Letters** literature, erudition
149 **use of service** keeping servants
  succession inheritance
150 **Bourn, bound of land** These are synonyms, like 'foison' and 'abundance', l. 161.
  tilth raising crops
151 **corn . . . oil** Noble (p. 80) points to an echo of the version of Psalm 4: 8 in the Psalter of the *Book of Common Prayer* :' . . . since the time that their corn, and wine, and oil, increased.' (p. 80).
152–3 **idle . . . but innocent and pure** countering the proverb 'Idleness begets lust' (Tilley 19)
157 **in common** for communal use
158 **Without sweat** The prelapsarian qualities implied in Montaigne become explicit in Gonzalo's commonwealth: compare Genesis 3: 19.

Sword, pike, knife, gun, or need of any engine
Would I not have, but nature should bring forth   160
Of it own kind all foison, all abundance
To feed my innocent people.

SEBASTIAN
No marrying 'mong his subjects?

ANTONIO
None, man, all idle—whores and knaves.

GONZALO
I would with such perfection govern, sir,
T'excel the golden age.

SEBASTIAN 'Save his majesty!

ANTONIO Long live Gonzalo!

GONZALO And—do you mark me, sir?

ALONSO
Prithee no more. Thou dost talk nothing to me.   170

GONZALO I do well believe your highness, and did it to minister occasion to these gentlemen, who are of such sensible
and nimble lungs that they always use to laugh at nothing.

ANTONIO 'Twas you we laughed at.

GONZALO Who in this kind of merry fooling am nothing to
you; so you may continue, and laugh at nothing still.

ANTONIO What a blow was there given!

SEBASTIAN An it had not fall'n flat-long.

GONZALO You are gentlemen of brave mettle; you would lift   180
the moon out of her sphere if she would continue in it five
weeks without changing.

Enter Ariel invisible, playing solemn music

182.1 *invisible*] MALONE; *not in* F

159 **engine** 'A machine or instrument used
in warfare' (*OED* 5a)
161 **it** its
foison plenty, abundance, specifically a
'plentiful crop or harvest' (*OED* 1b)
163 **No marrying** presumably not: marriage
is a 'contract' (l. 149), and irrelevant to
'innocent people' (l. 162).
167 **'Save** for 'God save', possibly shortened
in deference to the statute forbidding
oaths. Kermode observes that 'Jonson, in
revising *Every Man In His Humour*, altered
"God save" to "save" in 1.2.1' (p. 168).
171–72 **minister occasion** provide an

opportunity (to laugh)
172 **sensible** sensitive
173 **use** are accustomed
179 **An** if
flat-long on the flat of the sword, hence
harmlessly
180 **mettle** the same word as *metal* (F's spelling is 'mettal'), continuing Sebastian's
sword metaphor
180–82 **you would lift . . . changing** (a) You
would try to steal the moon if it held still
long enough; (b) the moon would have to
stop changing before you would do anything extraordinary. The retort probably

SEBASTIAN We would so, and then go a-bat-fowling.

ANTONIO Nay, good my lord, be not angry.

GONZALO No, I warrant you, I will not adventure my
discretion so weakly. Will you laugh me asleep, for I am
very heavy?

ANTONIO Go sleep, and hear us.

All sleep except Alonso, Sebastian, and Antonio

ALONSO
What, all so soon asleep? I wish mine eyes
Would, with themselves, shut up my thoughts. I find   190
They are inclined to do so.

SEBASTIAN   Please you, sir,
Do not omit the heavy offer of it.
It seldom visits sorrow; when it doth,
It is a comforter.

ANTONIO   We two, my lord,
Will guard your person while you take your rest,
And watch your safety.

ALONSO   Thank you. Wondrous heavy.

Alonso sleeps. Exit Ariel

SEBASTIAN
What a strange drowsiness possesses them!

ANTONIO
It is the quality o'th' climate.

SEBASTIAN   Why
Doth it not then our eyelids sink? I find not
Myself disposed to sleep.

ANTONIO   Nor I; my spirits are nimble.   200
They fell together all, as by consent;

188.1 *All . . . Antonio*] CAMBRIDGE; *Gon⟨zalo⟩., Adr⟨ian⟩., Fra⟨ncisco⟩. and Train, sleep* CAPELL;
not in F   196.1 *Alonso sleeps*] CAPELL, *not in* F   *Exit Ariel*] MALONE; *not in* F   201 consent:]
CAPELL; consent F; consent, POPE

alludes to the proverb 'The moon keeps
her course for all the dogs' barking': Gonzalo sneers at the two courtiers' ar-
rogance and ineffectuality.

183 **a-bat-fowling** (a) 'the catching of birds
by night when at roost' (*OED*), here using
the moon as a lantern; (b) 'swindling,
victimizing the simple' (*OED* 2).

185–6 **adventure my discretion so weakly**
put my good judgement at risk by such
weak behaviour

187 **heavy** sleepy

188 **Go . . . as** 'Compose yourself for sleep,
and we will do our part by laughing' (Kermode).

192 **omit** disregard (*OED* 2c)
heavy here including the sense 'serious'

198–9 **Why | Doth it not then our eye-
lids sink?** Prospero's 'project'; compare
l. 297. See the Introduction, pp. 53–4.

201 **consent** common agreement, consensus

They dropped as by a thunder-stroke. What might,
Worthy Sebastian, O what might—? No more.
And yet methinks I see it in thy face,
What thou shouldst be. Th' occasion speaks thee, and
My strong imagination sees a crown
Dropping upon thy head.

SEBASTIAN   What? Art thou waking?

ANTONIO
Do you not hear me speak?

SEBASTIAN   I do, and surely
It is a sleepy language, and thou speak'st
Out of thy sleep. What is it thou didst say?   210
This is a strange repose, to be asleep
With eyes wide open—standing, speaking, moving,
And yet so fast asleep.

ANTONIO   Noble Sebastian,
Thou let'st thy fortune sleep—die, rather; wink'st
Whiles thou art waking.

SEBASTIAN   Thou dost snore distinctly.
There's meaning in thy snores.

ANTONIO
I am more serious than my custom. You
Must be so too, if heed me; which to do
Trebles thee o'er.

SEBASTIAN   Well? I am standing water.

ANTONIO
I'll teach you how to flow.

---

205 **Th'occasion speaks thee** the opportunity proclaims to thee. Coleridge (cited in
the Variorum, p. 111) and Wilson Knight
(*The Crown of Life* (Oxford, 1947),
pp. 212–13) drew attention to analogies
between this scene and *Macbeth* Act 1,
Scene 3.
207 **waking** awake
214 **wink'st** you close your eyes (to this opportunity)
215 **distinctly** articulately, 'so as to be clearly perceived or understood' (*OED* 2)
219 **Trebles thee o'er** makes thee three times
greater. Sisson suggests a metaphor from
draughts: to treble is to jump three pieces
in one move (*New Readings*, i. 48).

219–20 **I am standing . . . ebb** Dent (E56.1)
cites 'Neither ebb nor flow, but just standing waters', with analogues from Webster
and Overbury, but doubts that the expression is proverbial. In any case, Sebastian contradicts it at ll. 220–21. M. S.
Luria (*ES* 49 (1968), 328–31) and J. E.
M. Latham (*N. & Q.* 21 (1974), 136)
draw attention to a traditional association
of sloth with standing water; Latham
cites a convincing parallel from Hall's
Characters (1608), 'Of the Slothful'.
219 **standing water** i.e. waiting to be moved
220–21 **to ebb . . . me** (a) My natural lazi-
ness prompts me to withdraw; (b) The
idleness imposed on me by my birth (i.e.

SEBASTIAN   Do so—to ebb   220
  Hereditary sloth instructs me.
ANTONIO   O!
  If you but knew how you the purpose cherish
  Whiles thus you mock it, how in stripping it
  You more invest it—ebbing men, indeed,
  Most often do so near the bottom run
  By their own fear or sloth.
SEBASTIAN   Prithee say on.
  The setting of thine eye and cheek proclaim
  A matter from thee, and a birth, indeed,
  Which throes thee much to yield.
ANTONIO   Thus, sir:
  Although this lord of weak remembrance, this,   230
  Who shall be of as little memory
  When he is earthed, hath here almost persuaded—
  For he's a spirit of persuasion, only
  Professes to persuade—the King his son's alive,
  'Tis as impossible that he's undrowned
  As he that sleeps here swims.
SEBASTIAN   I have no hope
  That he's undrowned.
ANTONIO   O, out of that no hope
  What great hope have you! No hope that way is
  Another way so high a hope that even
  Ambition cannot pierce a wink beyond,   240

by being Alonso's younger brother)
teaches me to hold back. 'Antonio's reply
dwells on the implications of (b), and ends
with a reproach aimed at the indolence
confessed in (a)' (Kermode).
222–3 **If . . . mock it** 'If you only understood
  your true feelings, realized that your
  mockery reveals how great your desire is.'
223–4 **in stripping it | You more invest it**
  The more you put it off the more important it becomes to you. *Invest* suggests a
  ceremonial robing.
227 **setting** fixed expression, 'set'
228 **A matter** something important
229 **throes thee much to yield** gives you
  much pain to produce. *Throe* as a verb is
  rare, but compare *Antony* 3.7.80: 'With
  news the time's in labour, and throes

forth | Each minute some.'
230 **of weak remembrance** whose memory
  is weak (alluding presumably to l. 155)
232 **earthed** buried
233 **spirit of persuasion** quintessentially a
  persuader
233–4 **only | Professes to persuade** giving
  counsel is his sole profession
240–1 **Ambition . . . there** The sense is that
  ambition cannot conceive of anything
  higher than the hope of a crown, but the
  syntax is confused, and when Antonio
  says that 'ambition cannot *but* doubt
  discovery' he is in fact saying the opposite
  of what the meaning requires. Furness
  suggested emending 'doubt' to 'douts'
  (puts out or extinguishes a fire), and
  glossed the passage, 'When ambition has

But doubt discovery there. Will you grant with me
That Ferdinand is drowned?
SEBASTIAN   He's gone.
ANTONIO   Then tell me,
Who's the next heir of Naples?
SEBASTIAN   Claribel.
ANTONIO
She that is Queen of Tunis; she that dwells
Ten leagues beyond man's life; she that from Naples
Can have no note unless the sun were post—
The man i' th' moon's too slow— till newborn chins
Be rough and razorable; she that from whom
We all were sea-swallowed, though some cast again—
And by that destiny, to perform an act   250
Whereof what's past is prologue, what to come
In yours and my discharge.
SEBASTIAN   What stuff is this? How say you?
'Tis true my brother's daughter's Queen of Tunis,
So is she heir of Naples, 'twixt which regions
There is some space.

---

250 And . . . destiny] ROWE; (And by that destiny) F; And that by destiny JOHNSON

pierced to its furthest wink, there discovery ceases and the crown is found'.
The only editor to accept this reading has
been Dover Wilson; it strikes me as far-
fetched, though it is possible that a play
• on doubt/dout is involved. The simplest
explanation of the difficulty may be the
correct one: the confusion of Antonio's
syntax expresses the anxiety and conflict
of his mind.
240 **pierce a wink** catch a glimpse
245 **Ten leagues beyond man's life** i.e.
farther than a man can travel in a
lifetime. A league was a variable measure,
usually about 3 miles, 'never in regular
use in England, but often occurring in
poetical or rhetorical statements of distance' (*OED*). Compare Antonio's use of
'cubit', l. 255. The actual distance from
Tunis to Naples is about 300 miles.
246 **note** information
    post messenger
247 **moon's too slow** The point is that the
moon requires a month to complete its
cycle, whereas the sun takes only a day.

248 **from** coming from
249 **cast** cast up (on shore); the word is also
apparently related to the theatrical
metaphor of ll. 250–1. *Cast* in the sense of
'assign parts in a play' is not recorded by
the *OED* before the eighteenth century,
but a *cast* as a theatrical role appears in
Euphues (*OED* 2b), and the subtitle of
Braithwait's *Whimzies* (1631) is *A New
Cast of Characters*. The word is not
elsewhere used by Shakespeare in a
theatrical sense.
250 **And by that destiny** F's parentheses
make Johnson's emendation, 'And that
by destiny', attractive. But Antonio's argument is not merely that they were cast
up by destiny, but that destiny has also
singled them out to perform the murders
he is proposing.
252 **discharge** performance, including a
continuation of the stage metaphor: compare Bottom's 'I will discharge it [the role
of Pyramus] in either your straw-colour
beard, your orange-tawny beard . . .'
(*Dream* 1.2.84 ff.).

ANTONIO A space whose every cubit
Seems to cry out, 'How shall that Claribel
Measure us back to Naples? Keep in Tunis,
And let Sebastian wake.' Say this were death
That now hath seized them, why, they were no worse
Than now they are. There be that can rule Naples   260
As well as he that sleeps, lords that can prate
As amply and unnecessarily
As this Gonzalo; I myself could make
A chough of as deep chat. O, that you bore
The mind that I do, what a sleep were this
For your advancement! Do you understand me?

SEBASTIAN
Methinks I do.

ANTONIO   And how does your content
Tender your own good fortune?

SEBASTIAN   I remember
You did supplant your brother Prospero.

ANTONIO   True;
And look how well my garments sit upon me,   270
Much feater than before. My brother's servants
Were then my fellows, now they are my men.

SEBASTIAN But for your conscience?

ANTONIO

Ay, sir, where lies that? If 'twere a kibe
'Twould put me to my slipper, but I feel not
This deity in my bosom. Twenty consciences

---

255 **cubit** a measure originally derived from the length of the forearm, 'varying at different times and places, but usually about 18–22 inches' (*OED* 2). Compare 'league', l. 245.
257 **Measure us** traverse us (cubits)
Keep stay
258 **wake** Compare ll. 214–15.
263–4 **I . . . chat** I could teach a jackdaw to speak as profoundly as he does. *Chough*: 'a bird of the crow family, formerly applied somewhat widely to all the smaller chattering species, but especially to the common jackdaw' (*OED*). For the figurative use, a chatterer or prater, the *OED* cites only this passage. In *All's Well*, the gibberish spoken to Parolles is called 'chough's language' (4.1.27), and Middleton and Rowley, in *A Fair Quarrel*

(1617), used the word for the name of a pretentious boor who comes to London from Cornwall to learn 'roaring' and fashionable speech (e.g. 'If thou wilt have it in plain terms, she is a callicut and a panagron' (5.1.159)). The word is also a variant spelling of *chuff*, a boor or churl, which provides a relevant ambiguity here.
267 **content** liking
268 **Tender** regard with either favour or fear (*OED* s.v. 2, 3b, e). The question is, are you inclined to look favourably on your good fortune or not?
271 **feater** i.e. they suit me better
274–6 **If . . . bosom** If it were a chilblain (kibe) on my heel it would force me to wear a slipper, but conscience causes me no inner discomfort.

That stand 'twixt me and Milan, candied be they,
And melt ere they molest! Here lies your brother,
No better than the earth he lies upon,
If he were that which now he's like—that's dead—   280
Whom I with this obedient steel, three inches of it,
Can lay to bed for ever; whiles you, doing thus,
To the perpetual wink for aye might put
This ancient morsel, this Sir Prudence, who
Should not upbraid our course. For all the rest,
They'll take suggestion as a cat laps milk;
They'll tell the clock to any business that
We say befits the hour.

SEBASTIAN   Thy case, dear friend,
Shall be my precedent: as thou got'st Milan,
I'll come by Naples. Draw thy sword—one stroke   290
Shall free thee from the tribute which thou payest,
And I the King shall love thee.

ANTONIO   Draw together,
And when I rear my hand do you the like
To fall it on Gonzalo.

SEBASTIAN   O, but one word.
          They talk apart.
          Enter Ariel, invisible, with music and song

ARIEL
    My master through his art foresees the danger

294.1 *They talk apart*] CAPELL; *not in* F   294.2 *invisible*] CAPELL; *not in* F

277 **candied** The sense is probably 'con-
  gealed, frozen solid', rather than 'turned
  to sugar, glazed'.
278 **melt** possibly an apocope for 'melted':
  see Abbott 472.
  molest interfere with me
282 **doing thus** Antonio mimes stabbing
  Gonzalo.
283 **To . . . put** might put to sleep forever
284 **morsel** choice dish; Kittredge suggests
  'fragment of a man'. For the playful or
  sarcastic application to a person, compare
  Measure 3.2.56, 'How doth my dear morsel, thy mistress?' and *K. John* 4.3.143,
  'this morsel of dead royalty'.
286 **suggestion** 'prompting or incitement to
  evil' (*OED* 1)
  as . . . milk i.e. naturally and eagerly

287–8 **tell . . . hour** say it is time to do whatever business we say is apropos
294.2 **invisible, with music** Ariel is no
  longer dressed as a water nymph; *invisible*
  implies nothing about his costume, but
  only that the other characters cannot see
  him. *With music* may mean that he carries
  an instrument on which to accompany
  himself in the song, as he did at 1.2.374,
  or it may mean that he is accompanied by
  a musical consort—compare the music
  that Ferdinand hears 'above me',
  1.2.408.
295–7 **My master . . . living** Ariel acts briefly
  as a chorus. The lines are addressed not to
  the sleeping Gonzalo but to the audience
  (the song that wakes Gonzalo is perceived
  by him only as 'a humming', l. 315).

That you, his friend, are in, and sends me forth—
For else his project dies—to keep them living.
    He sings in Gonzalo's ear
        While you here do snoring lie,
        Open-eyed conspiracy
          His time doth take.   300
        If of life you keep a care,
        Shake off slumber, and beware.
        Awake, awake!

ANTONIO
Then let us both be sudden.
GONZALO (*waking*)   Now, good angels
  Preserve the King!
    The others wake

ALONSO
Why, how now, ho! Awake? Why are you drawn?
Wherefore this ghastly looking?
GONZALO   What's the matter?

SEBASTIAN
Whiles we stood here securing your repose,
Even now, we heard a hollow burst of bellowing,
Like bulls, or rather lions—did't not wake you?   310
It struck mine ear most terribly.
ALONSO   I heard nothing.

ANTONIO
O, 'twas a din to fright a monster's ear,
To make an earthquake. Sure, it was the roar
Of a whole herd of lions.
ALONSO   Heard you this, Gonzalo?

GONZALO
Upon mine honour, sir, I heard a humming,
And that a strange one too, which did awake me.
I shaked you, sir, and cried. As mine eyes opened

---

304 *waking*] DYCE; *not in* F   305.1 *The others wake*] CAMBRIDGE; *They wake* ROWE; *not in* F

297 **project** The word combines the meanings of both scheme and purpose (see
  5.1.1). For Prospero's project here, see
  the Introduction, pp. 53–4.

them Gonzalo and Alonso
307 **ghastly** full of fear (*OED* 3)
308 **securing** guarding

143

I saw their weapons drawn. There was a noise,
That's verily. 'Tis best we stand upon our guard,
Or that we quit this place. Let's draw our weapons.   320
ALONSO
Lead off this ground, and let's make further search
For my poor son.
GONZALO   Heavens keep him from these beasts!
For he is sure i' th' island.
ALONSO   Lead away.
ARIEL
Prospero my lord shall know what I have done.
So, King, go safely on to seek thy son.   *Exeunt*

2.2   *Enter Caliban with a burden of wood*
CALIBAN
All the infections that the sun sucks up
From bogs, fens, flats, on Prosper fall, and make him
By inchmeal a disease!
          A noise of thunder heard
                              His spirits hear me,
And yet I needs must curse. But they'll nor pinch,
Fright me with urchin-shows, pitch me i' th' mire,
Nor lead me like a firebrand in the dark
Out of my way, unless he bid 'em; but
For every trifle are they set upon me,
Sometime like apes that mow and chatter at me,
And after bite me; then like hedgehogs, which   10
Lie tumbling in my barefoot way, and mount
Their pricks at my footfall; sometime am I
All wound with adders, who with cloven tongues

2.2.0.1 *Enter . . . wood*] F *adds 'A noise of thunder heard.'*   3 *A . . . heard*] *in* F *at* 0.1

---

325 *Exeunt* Ariel probably departs in a different direction from the rest.
2.2.2 **flats** swamps
  3 **By inchmeal** inch by inch
  3 *A noise of thunder heard* F has this in
parentheses as part of the opening stage
direction, but it seems more likely to
belong here: Caliban takes the thunder as
a threatening response to his curse.
  5 **urchin-shows** goblin shows; apparitions

in the shape of hedgehogs (compare
1.2.326–8 and 2.2.10)
  6 **firebrand** literally a piece of wood kindled
at the fire; not recorded as a term for a
will-o'-the-wisp. Shakespeare's use is
metaphorical.
  9 **mow** make mouths or grimaces
 10 **like hedgehogs** the *urchin-shows* of l. 5
 13 **wound with** entwined by

Do hiss me into madness—
    Enter Trinculo   Lo, now, lo,
Here comes a spirit of his, and to torment me
For bringing wood in slowly. I'll fall flat.
Perchance he will not mind me.
    He lies down and covers himself with his cloak

TRINCULO Here's neither bush nor shrub to bear off any
weather at all, and another storm brewing—I hear it
sing i' th' wind. Yon same black cloud, yon huge one, 20
looks like a foul bombard that would shed his liquor. If it
should thunder as it did before, I know not where to hide
my head—yon same cloud cannot choose but fall by
pailfuls. What have we here—a man or a fish?—dead
or alive? A fish, he smells like a fish; a very ancient and
fish-like smell; a kind of not-of-the-newest poor-John. A
strange fish! Were I in England now, as once I was, and
had but this fish painted, not a holiday-fool there but
would give a piece of silver. There would this monster
make a man—any strange beast there makes a man. 30
When they will not give a doit to relieve a lame beggar,
they will lay out ten to see a dead Indian. Legged like a
man, and his fins like arms! Warm, o'my troth! I do now
let loose my opinion, hold it no longer: this is no fish, but
an islander, that hath lately suffered by a thunderbolt.
(*Thunder*) Alas, the storm is come again! My best way is
to creep under his gaberdine—there is no other shelter

17.1 *He . . . cloak*] This edition; *not in* F   36 *Thunder*] CAPELL; *not in* F

14 **Trinculo** The name is related to Italian *trincare*, drink deeply, and *trincone*, a drunkard.
17 **mind** notice
18 **bear off** keep off
21 **bombard** the earliest kind of cannon, and thence 'a leather jug or bottle for liquor . . . probably from some resemblance to the early cannons' (*OED* 3). The connotation of offensive weaponry is still clearly present in Shakespeare's usage.
26 **poor-John** dried, salted fish; poor food
28 **had . . . painted** on a sign, to attract spectators
30 **make a man** both 'make a man's fortune'

and 'be considered a man'
31 **doit** a coin of small value, originally equivalent to half a farthing
32 **a dead Indian** The display of New–World natives, living and dead, had been a popular and lucrative enterprise since the early sixteenth century. Sir Martin Frobisher brought back and exhibited live Indians in 1576 and 1577. 'Such exhibitions were profitable investments, and were a regular feature of colonial policy under James I. The exhibits rarely survived the experience . . .' (Kermode).
35 **suffered** perished
37 **gaberdine** cloak of coarse cloth

hereabout. Misery acquaints a man with strange bedfellows. I will here shroud till the dregs of the storm be
past.   40

    He crawls under Caliban's cloak.

    Enter Stephano singing, a bottle in his hand

STEPHANO

             I shall no more to sea, to sea,
             Here shall I die ashore—
This is a very scurvy tune to sing at a man's funeral.
Well, here's my comfort. (*Drinks*)
(*Sings*)

             The master, the swabber, the boatswain, and I,
               The gunner, and his mate,
             Loved Moll, Meg, and Marian, and Margery,
               But none of us cared for Kate;
             For she had a tongue with a tang,
             Would cry to a sailor, 'Go hang!'   50
       She loved not the savour of tar nor of pitch,
       Yet a tailor might scratch her where'er she did itch.
             Then to sea, boys, and let her go hang!
This is a scurvy tune too, but here's my comfort.

    He drinks

CALIBAN Do not torment me! O!

STEPHANO What's the matter? Have we devils here? Do you
put tricks upon's with savages and men of Ind? Ha? I
have not scaped drowning to be afeard now of your four

40.1 *He . . . cloak*] This edition; *not in* F   40.2 *a . . . hand*] CAPELL; *not in* F

38–9 **Misery . . . bedfellows** not recorded as
a proverb
39 **dregs** continuing the 'bombard' metaphor of l. 21
45 **swabber** seaman who cleans the decks
49 **tang** sting; originally 'the tongue of a
serpent, formerly thought to be a stinging
organ' (*OED*).
52 **tailor** Tailors were conventionally supposed to be unmanly.
    scratch . . . itch implying the gratification
of sensual desire.
56 **What's the matter?** What's going on?
56–7 **Do . . . upon's** 'To put a trick upon'
someone was proverbial (Dent PP18).
Tilley cites a 1709 proverb, 'Do not put
tricks upon travellers' (T521), but there is

no reason to believe that this predates *The
Tempest*.
57 **savages** F's 'salvages' was a common
variant spelling and is the form regularly
used by Ralph Crane. It carries with it no
special implications. Kermode's belief that
it relates Caliban to the traditional figure
of the wodewose or European wild man is
discussed in the Introduction, pp. 25–6.
men of Ind. (West) Indian natives; the
phrase is in apposition with *savages*. Noble
(p. 250) notes that the expression appears
in Jeremiah 13: 23 in the Bishops' Bible:
'May a man of Ind change his skin . . . ?'
(The Authorized Version reads 'Ethio-
pian'.) For another possible allusion to
this verse, see 4.1.262.

legs; for it hath been said, 'As proper a man as ever went
on four legs cannot make him give ground'; and it shall   60
be said so again, while Stephano breathes at' nostrils.

CALIBAN The spirit torments me! O!

STEPHANO This is some monster of the isle with four legs,
who hath got, as I take it, an ague. Where the devil
should he learn our language? I will give him some relief,
if it be but for that. If I can recover him, and keep him
tame, and get to Naples with him, he's a present for any
emperor that ever trod on neat's-leather.

CALIBAN Do not torment me, prithee! I'll bring my wood
home faster.   70

STEPHANO He's in his fit now, and does not talk after the
wisest. He shall taste of my bottle. If he have never drunk
wine afore, it will go near to remove his fit. If I can recover
him and keep him tame, I will not take too much for him;
he shall pay for him that hath him, and that soundly.

CALIBAN Thou dost me yet but little hurt; thou wilt anon, I
know it by thy trembling. Now Prosper works upon thee.

STEPHANO Come on your ways. Open your mouth—here
is that which will give language to you, cat. Open your
mouth—this will shake your shaking, I can tell you,   80
and that soundly. (*Caliban drinks*) You cannot tell who's
your friend—open your chops again.

TRINCULO I should know that voice. It should be—but he is
drowned, and these are devils—O defend me!

STEPHANO Four legs and two voices; a most delicate monster! His forward voice now is to speak well of his friend,

---

81 *Caliban drinks*] This edition; *not in* F

59–60 **As . . . ground** The proverbial for-
mula is 'As good a man as ever went on
two legs' (Dent M66), but Stephano
adapts it to the four-legged monster he
sees.

61 **at' nostrils** With *at* 'the article is generally omitted in . . . adverbial forms'
(Abbott 143).

64 **ague** Commonly used for the shivering
stage of fever, hence any fit of shaking or
quaking, as Caliban is doing under his
gaberdine.

66 **recover** cure

68 **that . . . leather** proverbial (see Tilley
M66)
neat's-leather cowhide

71–2 **after the wisest** in the wisest fashion

73 **go near to** do much to

74 **I will not take too much for him** No price
will be too high for him.

77 **thy trembling** Trinculo is now quaking
with fear.

78–9 **here . . . cat** the proverbial 'liquor that
would make a cat speak' (Tilley A99)

81–2 **You . . . friend** Presumably Caliban
dislikes his first taste.

85 **delicate** exquisitely made

his backward voice is to utter foul speeches and to detract.
If all the wine in my bottle will recover him, I will help his
ague. Come. (*Caliban drinks again*) Amen! I will pour some
in thy other mouth.   90

TRINCULO Stephano!

STEPHANO Doth thy other mouth call me? Mercy, mercy!
This is a devil, and no monster. I will leave him; I have no
long spoon.

TRINCULO Stephano! If thou beest Stephano, touch me, and
speak to me; for I am Trinculo—be not afeard—thy good
friend Trinculo.

STEPHANO If thou beest Trinculo, come forth. I'll pull thee
by the lesser legs—if any be Trinculo's legs, these are
they. (*Pulls him from under the cloak*) Thou art very   100
Trinculo indeed! How cam'st thou to be the siege of this
mooncalf? Can he vent Trinculos?

TRINCULO I took him to be killed with a thunder-stroke. But
art thou not drowned, Stephano? I hope now thou art not
drowned. Is the storm overblown? I hid me under the
dead mooncalf's gaberdine for fear of the storm. And art
thou living, Stephano? O, Stephano, two Neapolitans
scaped!

STEPHANO Prithee do not turn me about; my stomach is not
constant.   110

CALIBAN (*aside*) These be fine things, an if they be not
sprites. That's a brave god, and bears celestial liquor. I
will kneel to him.

---

89 *Caliban drinks again*] WILSON; *not in* F   100 *Pulls . . . cloak*] This edition; *not in* F

---

88 **If . . . him** if it takes all the wine in my
bottle to cure him
92 **call me** speak my name (implying supernatural knowledge)
93–4 **I . . . spoon** The proverb is 'He must
have a long spoon that will eat with the
devil' (Tilley S771), also alluded to in
Errors 4.3.63 ff.
101 **siege** excrement (*OED* 3c)
102 **mooncalf** monstrosity (regarded as
having been produced by the influence of
the moon at its birth); also a 'born fool'
vent defecate
109 **turn me about** Presumably Stephano,

in his relief, is attempting a kind of dance
with Trinculo.
112 **sprites, celestial liquor** It is tempting to
see throughout this scene a continuing
play on 'spirits' as liquor produced by dis-
tillation. The term in this precise sense
seems not to have been in use until the
later seventeenth century, though *The Alchemist*, written shortly before *The Tempest*, would have familiarized audiences
with the general notion of spirits as dis-
tilled liquids of any sort. Condensation of
an essence from vapour was apparently

STEPHANO How didst thou scape? How cam'st thou hither?
Swear by this bottle how thou cam'st hither—I escaped
upon a butt of sack which the sailors heaved o'erboard
—by this bottle, which I made of the bark of a tree with
mine own hands since I was cast ashore.

CALIBAN I'll swear upon that bottle to be thy true subject,
for the liquor is not earthly.   120

STEPHANO Here; swear then how thou escaped'st.

TRINCULO Swum ashore, man, like a duck. I can swim like
a duck, I'll be sworn.

STEPHANO Here, kiss the book. (*He gives Trinculo the bottle*)
Though thou canst swim like a duck, thou art made like
a goose.

TRINCULO O Stephano, hast any more of this?

STEPHANO The whole butt, man. My cellar is in a rock by the
seaside, where my wine is hid. How now, mooncalf, how
does thine ague?   130

CALIBAN Hast thou not dropped from heaven?

STEPHANO Out o' th' moon, I do assure thee. I was the man
'i th' moon when time was.

CALIBAN I have seen thee in her, and I do adore thee. My
mistress showed me thee, and thy dog and thy bush.

STEPHANO Come, swear to that: kiss the book. I will furnish
it anon with new contents. Swear.

   Caliban drinks

---

124 *He . . . bottle*] NEILSON–HILL (*Passing the bottle*); *not in* F   137.1 *Caliban drinks*] COLLIER; *not
in* F

considered analogous to the manifestation of a spirit in the air.

112 **That's a brave god** Caliban's reaction to
Stephano parallels Miranda's to Ferdinand ('a thing divine') and Ferdinand's
to Miranda ('most sure, the goddess . . .'),
1.2.419 ff.

116 **sack** The term was applied to any of a
variety of white wines imported from
Spain and the Canary Islands.

121 **Here . . . escaped'st** Stephano ignores
Caliban until l. 129.

124 **kiss the book** alluding both to kissing
the Bible to confirm an oath and to the
proverbial 'Kiss the cup' (Tilley C909)

126 **goose** Referring to Trinculo's posture as
he drinks; but also, the term was a

byword for giddiness and unsteadiness on
the feet.

132–5 **man . . . bush** The man, according to
the folktale, was banished to the moon,
variously for stealing a bundle of kindling
(his thorn bush), or for gathering kindling
on the sabbath. Belief in the man in the
moon was energetically attacked by
reforming zealots in the fifteenth century,
and later by Puritans, but was defended,
e.g. by the Lollard Reginald Pecock
(c. 1450), as an example of harmless
superstition (*The Repressor of Over Much
Blaming the Clergy*, ed. C. Babington
(London, 1860), i. 155). Compare *Dream*
3.1.55–7.

133 **when time was** once upon a time

TRINCULO By this good light, this is a very shallow monster.
    I afeard of him? A very weak monster! The man i' th'
    moon? A most poor, credulous monster! Well drawn,   140
    monster, in good sooth!
CALIBAN I'll show thee every fertile inch o' th' island—
    and I will kiss thy foot. I prithee be my god.
TRINCULO By this light, a most perfidious and drunken mon-
    ster. When's god's asleep, he'll rob his bottle.
CALIBAN I'll kiss thy foot. I'll swear myself thy subject.
STEPHANO Come on, then, down and swear.
TRINCULO I shall laugh myself to death at this puppy-headed
    monster. A most scurvy monster! I could find in my heart
    to beat him—   150
STEPHANO Come, kiss.
TRINCULO —But that the poor monster's in drink. An
    abominable monster!
CALIBAN
    I'll show thee the best springs; I'll pluck thee berries;
    I'll fish for thee, and get thee wood enough.
    A plague upon the tyrant that I serve!
    I'll bear him no more sticks, but follow thee,
    Thou wondrous man.
TRINCULO A most ridiculous monster, to make a wonder of
    a poor drunkard!   160
CALIBAN
    I prithee let me bring thee where crabs grow,
    And I with my long nails will dig thee pig-nuts,
    Show thee a jay's nest, and instruct thee how

---

138 **this good light** i.e. the sun
140 **Well drawn** a good draft, well drunk
142 **I'll . . . island** as he had done for Pros-
    pero: compare 1.2.337 ff.
158–9 **wondrous . . . wonder** Caliban's re-
    action to Stephano again parallels Fer-
    dinand's to Miranda: compare 'O you
    wonder', 1.2.427.
161 **crabs** Shakespeare uses the word for
    both crabapples and crustaceans, and it
    has been invariably assumed, because of
    the verb *grow*, that the former is intended
    here: crabs would be expected to 'dwell'.
    But there are places where crabs may be
    said to *grow* (rock-pools for example), and
    the verb alone is not a conclusive argu-

ment in favour of crabapples. Moreover,
crabapples were not considered good to
eat—their sourness was proverbial (see
Shrew 2.1.230)—and Caliban may well
be promising Stephano shellfish instead.
162 **pig-nuts** edible tubers (*burnium flexuo-
    sum*), also called earth-nuts and earth
    chestnuts
163 **jay's nest** *Jays* were prized for their
    plumage, and the nests tend to be well
    hidden. Since everything else on
    Caliban's list is edible, however, he may
    be offering Stephano the eggs, though
    there is no record of these being con-
    sidered delicacies: Compare the *marmoset*,
    l. 164.

To snare the nimble marmoset. I'll bring thee
To clust'ring filberts, and sometimes I'll get thee
Young scamels from the rock. Wilt thou go with me?
STEPHANO I prithee now lead the way without any more
talking. Trinculo, the King and all our company else
being drowned, we will inherit here. (*To Caliban*) Here,
bear my bottle. Fellow Trinculo, we'll fill him by and by　　170
again.
CALIBAN ⌈*sings drunkenly*⌉ Farewell, master, farewell,
farewell!
TRINCULO A howling monster; a drunken monster!
CALIBAN No more dams I'll make for fish,
　　Nor fetch in firing
　　At requiring,
Nor scrape trenchering, nor wash dish:
　　'Ban, 'Ban, Ca-Caliban
Has a new master—get a new man!　　180
Freedom, high-day! High-day, freedom! Freedom, high-
　　day, freedom!
STEPHANO O brave monster! Lead the way!　　　　*Exeunt*

---

169 *To Caliban*] This edition; *not in* F

---

164 **marmoset** a small monkey, common as
a pet, but also said in Harcourt's *Voyage to
Guiana* to be edible (cited by Kermode)
165 **filberts** hazelnuts
166 **scamels** This creature has provoked
endless debate. The context requires a
crustacean, bird, or a fish of the sort
'frequenting rocks . . ., such as the black
goby or sea-gudgeon, the striped bass, the
wrasse, etc.' (*OED* s. rock-fish). The most
common emendation is 'sea-mels', a
variant of sea-mew or gull; but Shakespeare may be adapting—or misunder-
standing—a foreign word from the same
body of travel literature that supplied him
with Setebos. 'French and Italian accounts of Magellan's . . . circumnaviga-
tion . . . relate that the men, off Patagonia,
ate small fish described as "*fort scameux*"
[very scaly] and "*squame*".' *Squama* is
found in several sixteenth-century dic-
tionaries, as well as in Florio's *Queen*

Anna's New World of Words (1611)
(Charles Frey, '*The Tempest* and the New
World', *ShQ*, 30 (1979), 29–41; p. 33).
170 **bear my bottle** probably addressed to
Caliban, as servant and attendant. Trinculo has a bottle at 3.2.64, but by that
time there may be more than one involved: Trinculo laments the loss of 'our
bottles' at 4.1.207.
　him either Caliban, or conceivably the
bottle being personified
172 *sings drunkenly* The stage direction
may be misplaced, and properly apply
only to the song, 'No more dams'.
176 **firing** firewood
178 **trenchering** trenchers collectively (a
sense not recorded elsewhere)
179 **'Ban** abbreviating Caliban
180 **get a new man** addressed to the old master, Prospero
181 **high-day** holiday

3.1   *Enter Ferdinand bearing a log*

FERDINAND

There be some sports are painful, and their labour
Delight in them set off; some kinds of baseness
Are nobly undergone; and most poor matters
Point to rich ends. This my mean task
Would be as heavy to me, as odious, but
The mistress which I serve quickens what's dead,
And makes my labours pleasures. O, she is
Ten times more gentle than her father's crabbed,
And he's composed of harshness. I must remove
Some thousands of these logs and pile them up,   10
Upon a sore injunction. My sweet mistress
Weeps when she sees me work, and says such baseness
Had never like executor. I forget.
But these sweet thoughts do even refresh my labours,
Most busil'est when I do it.

Enter Miranda, and Prospero at a distance, unseen

3.1.2 set] F; sets ROWE   15 busil'est] RIVERSIDE; busie lest F; busilest KERMODE (busiliest *conj.*
Bulloch) (*see note*)   15.1 at . . . unseen ROWE; *not in* F

3.1.1–2 **their labour | Delight in them set off**
Either 'the effort they require shows to
advantage the pleasure we take in them',
or 'the pleasure they give takes away
their laborious aspects'. *Set off* = either
'adds lustre to' or 'removes': most editors
gloss 'compensates for', but this seems to
be a later usage. All editors following
Rowe emend 'set' to 'sets', but it is not
clear that Rowe is correct. The labour (or
the pleasure) of sports may be considered
a plural subject: see analogous examples
in Abbott 337.
3 **most poor** poorest
4 **mean** lowly
5 **but** except that
11 **sore** harsh
12–13 **such . . . executor** Such base labour
was never performed by one so noble.
13 **I forget** i.e. to work at my task
14–15 **these . . . do it** My thoughts of
Miranda are most active when I am
busiest at my work. Kermode's emendation seems the most likely reading for this
famous crux. The editorial tradition has

on the whole assumed that Ferdinand
means that his thoughts of Miranda are
busiest when he is least busy, and has
emended the line accordingly, e.g. to
'most busiest when idlest' (Spedding), or
'most busy, least when I do it' (Collier,
Alexander). But, as Kermode argues in a
detailed and persuasive note, this is
precisely the opposite of the sense
required: 'The trouble with "*Most busy
least*", as with "*busiest . . . idlest*", is that
it involves Ferdinand in a *non sequitur*:
"these thoughts refresh my labours; I am
busiest when not labouring"—although
he has just been saying that his labour is
a pleasure because of these thoughts.' He
therefore emends F's 'busie lest' to
'busilest', following a suggestion of
Bulloch's; the adverbial superlative is
unusual, but is paralleled in *Cymbeline*
4.2.206–7, '. . . to show what coast thy
sluggish crare | Might easiest harbour in',
as well as in Q1 of *Lear* 1.2.132, 'the
maidenlest star in the firmament', where
F reads 'maidenliest'.

MIRANDA   Alas, now pray you
Work not so hard. I would the lightning had
Burnt up those logs that you are enjoined to pile!
Pray set it down, and rest you. When this burns,
'Twill weep for having wearied you. My father
Is hard at study. Pray now, rest yourself;   20
He's safe for these three hours.
FERDINAND   O most dear mistress,
The sun will set before I shall discharge
What I must strive to do.
MIRANDA   If you'll sit down
I'll bear your logs the while. Pray give me that;
I'll carry it to the pile.
FERDINAND   No, precious creature,
I had rather crack my sinews, break my back,
Than you should such dishonour undergo
While I sit lazy by.
MIRANDA   It would become me
As well as it does you, and I should do it
With much more ease, for my good will is to it,   30
And yours it is against.
PROSPERO (*aside*)   Poor worm, thou art infected!
This visitation shows it.
MIRANDA   You look wearily.
FERDINAND

No, noble mistress, 'tis fresh morning with me
When you are by at night. I do beseech you—
Chiefly that I might set it in my prayers—
What is your name?
MIRANDA   Miranda.—O my father,
I have broke your hest to say so.
FERDINAND   Admired Miranda,

19 **'Twill weep** by exuding resin
21 **He's safe** i.e. we are safe from him
31 **worm** any small creature; used variously
    to express affection, as here, or contempt;
    also, a source of infection: compare the
    effects of the 'worm in the bud' in *Twelfth
    Night* 2.4.111.
32 **visitation** punning both on the sense of
    'plague' (*OED* 2.6c) and on the pastoral
    or charitable visit to the sick (compare

2.1.12). Olivia after her first meeting with
Cesario/Viola similarly views love as an
infection: 'Even so quickly may one catch
the plague?' (*Twelfth Night* 1.5.290).
34 **When . . . night** The language is purely
    conventional: Ferdinand has not yet
    passed a night since meeting Miranda.
37 **hest** command
    Admired Miranda punning on her name

Indeed the top of admiration, worth
What's dearest to the world! Full many a lady
I have eyed with best regard, and many a time   40
Th' harmony of their tongues hath into bondage
Brought my too diligent ear. For several virtues
Have I liked several women, never any
With so full soul but some defect in her
Did quarrel with the noblest grace she owed,
And put it to the foil. But you, O you,
So perfect and so peerless, are created
Of every creature's best.

MIRANDA   I do not know
One of my sex, no woman's face remember,
Save from my glass, mine own; nor have I seen   50
More that I may call men than you, good friend,
And my dear father. How features are abroad
I am skilless of; but by my modesty,
The jewel in my dower, I would not wish
Any companion in the world but you;
Nor can imagination form a shape
Besides yourself to like of. But I prattle
Something too wildly, and my father's precepts
I therein do forget.

FERDINAND   I am, in my condition,
A prince, Miranda; I do think a king—   60
I would not so!—and would no more endure
This wooden slavery than to suffer

---

62 wooden] F2 (woodden); wodden F1

39 **dearest** most valuable
42 **diligent** attentive (*OED* 3)
42, 43 **several** various, different
45 **owed** owned
46 **put it to the foil** either foiled it, overthrew
it (*OED* sb²), or, taken with *quarrel* (l. 45),
challenged it, as at a fencing match. The
alternative meaning, cited by most
editors, 'set it off to advantage by
contrast' (*OED* 6), gives the opposite of
the required sense.
47–8 **So . . . best** 'Alluding to the picture of
Venus by Apelles' (Johnson): this was a
synthesis of the most perfect features of
the most beautiful women the painter

could find. Steevens cited a fable from the
Third Eclogues of *Arcadia* III, in which
man is created out of a composite of the
salient qualities of every animal. Compare
Orlando's poem to Rosalind, *As You Like It*
3.2.139 ff.
48 **creature** created being
52–3 **How . . . of** I do not know what people
look like elsewhere.
53 **skilless** ignorant
    modesty virginity
54 **dower** (here) dowry
59 **condition** rank
61 **would not** wish it were not

The flesh-fly blow my mouth. Hear my soul speak:
The very instant that I saw you did
My heart fly to your service, there resides
To make me slave to it, and for your sake
Am I this patient log-man.

MIRANDA   Do you love me?

FERDINAND
O heaven, O earth, bear witness to this sound,
And crown what I profess with kind event
If I speak true; if hollowly, invert   70
What best is boded me to mischief: I,
Beyond all limit of what else i' th' world,
Do love, prize, honour you.

MIRANDA   I am a fool
To weep at what I am glad of.

PROSPERO *(aside)*   Fair encounter
Of two most rare affections! Heavens rain grace
On that which breeds between 'em!

FERDINAND   Wherefore weep you?

MIRANDA
At mine unworthiness, that dare not offer
What I desire to give, and much less take
What I shall die to want. But this is trifling,
And all the more it seeks to hide itself,   80
The bigger bulk it shows. Hence, bashful cunning,
And prompt me, plain and holy innocence!
I am your wife if you will marry me;
If not, I'll die your maid. To be your fellow
You may deny me, but I'll be your servant
Whether you will or no.

---

63 **flesh-fly** fly that deposits its eggs in dead
flesh
blow corrupt ('said of flies and other in-
sects: to deposit eggs' (*OED* iii. 28))
69 **event** outcome
70 **hollowly** insincerely, falsely (*OED* s.
hollow, 5)
70–1 **invert . . . mischief** turn whatever
good fortune is foretold for me to evil
72 **what** whatever (Abbott 254)
77–81 **that dare not offer . . . cunning** The

offer is unconsciously sexual; with *this is
trifling* Miranda realizes her indiscretion,
acknowledges the *bashful cunning* (81) of
her language, and undertakes to confront
its implications.
79 **want** lack
80–1 **all . . . shows** 'The imagery here is of a
secret pregnancy' (Barton).
81 **bashful cunning** dissimulating bashful-
ness
84 **maid** (a) virgin; (b) servant (see l. 85)

FERDINAND   My mistress, dearest,
And I thus humble ever.
    He kneels
MIRANDA   My husband then?
FERDINAND   Ay, with a heart as willing
As bondage e'er of freedom. Here's my hand.
MIRANDA
And mine, with my heart in't. And now farewell   90
Till half an hour hence.
FERDINAND   A thousand-thousand!
    Exeunt Ferdinand and Miranda separately
PROSPERO
So glad of this as they I cannot be,
Who are surprised withal, but my rejoicing
At nothing can be more. I'll to my book,
For yet ere suppertime must I perform
Much business appertaining.   *Exit*

3.2   *Enter Caliban, Stephano, and Trinculo*
STEPHANO (*to Trinculo*) Tell not me. When the butt is out, we
   will drink water; not a drop before. Therefore bear up and
   board 'em: servant-monster, drink to me.
TRINCULO Servant-monster! The folly of this island! They
   say there's but five upon this isle: we are three of them;

87.1 *He kneels*] This edition; *not in* F   91 thousand-thousand] This edition; thousand,
thousand F (*see note*)   91.1 *Exeunt . . . separately*] CAPELL (*Exeunt . . . severally*); *Exeunt.* F
93 withal] THEOBALD; with all F
  3.2.1 *to Trinculo*] This edition; *not in* F

86 **mistress** 'A woman who has command
over a man's heart' (*OED* 10), in this context without illicit overtones
88 **willing** desirous (*OED* 1)
89–90 **my hand . . . in't** 'With heart and
hand' is proverbial (Tilley H339).
91 **thousand-thousand** i.e. a million, as in
Twelfth Night 2.4.63. F prints 'thousand,
thousand': for the comma with the force
of a hyphen, compare 'marriage-blessing'
(F's 'marriage, blessing') 4.1.106.
93 **surprised** both astonished and caught
unawares
  withal '"This" is understood after *withal*,
so that it means "with all this", and is
used adverbially' (Abbott 196).
3.2.1 **Tell not me** Trinculo has been trying to

moderate their drinking.
  butt is out cask is finished
2 **bear up** sail up (to an enemy ship): the
naval order here means 'drink up'.
3 **servant-monster** The epithet is used by Ben
Jonson in *Bartholomew Fair* (1614), apparently as part of an invidious comparison
with Shakespeare: 'If there be never a servant-monster i' the fair, who can help it? . . .
nor a nest of antics? [The author] is loth to
make nature afraid in his plays, like those
that beget tales, tempests, and such-like
drolleries . . .' (Induction, ll. 127–30).
4 **folly** absurdity (*OED* 1c), and the foolishness alluded to in ll. 5–6. Kermode suggests 'freak', citing Evelyn, and that the
reference is to Caliban.

if th'other two be brained like us, the state totters.

STEPHANO Drink, servant-monster, when I bid thee. Thy eyes are almost set in thy head.

TRINCULO Where should they be set else? He were a brave monster indeed if they were set in his tail! 10

STEPHANO My man-monster hath drowned his tongue in sack. For my part, the sea cannot drown me: I swam, ere I could recover the shore, five and thirty leagues off and on, by this light. Thou shalt be my lieutenant-monster, or my standard.

TRINCULO Your lieutenant if you list; he's no standard.

STEPHANO We'll not run, Monsieur Monster.

TRINCULO Nor go neither; but you'll lie like dogs, and yet say nothing neither.

STEPHANO Mooncalf, speak once in thy life, if thou beest a 20
good mooncalf.

CALIBAN How does thy honour? Let me lick thy shoe. I'll not serve him, he is not valiant.

TRINCULO Thou liest, most ignorant monster: I am in case to jostle a constable. Why thou debauched fish, thou, was there ever man a coward that hath drunk so much sack as I today? Wilt thou tell a monstrous lie, being but half a fish and half a monster?

CALIBAN Lo, how he mocks me! Wilt thou let him, my lord?

TRINCULO 'Lord', quoth he? That a monster should be such 30
a natural!

CALIBAN Lo, lo again! Bite him to death, I prithee.

---

14 on, . . . Thou] CAPELL; on, by this light thou F   25 debauched] F (deboshed)

---

6 **be brained like us** i.e. are no more intelligent than we are
8 **set** fixed (drunkenly)
13 **five and thirty leagues** A league was a variable measure, usually about 3 miles; hence Stephano is claiming to have swum over 100 miles.
13–14 **off and on** either one way and another, or intermittently
off . . . light The Folio attaches the oath to Stephano's determination to promote Caliban; Capell's emendation attaches it to Stephano's lie. Both require minor repointing, and Capell's gives the stronger dramatic sense.
15 **standard** standard-bearer

16 **he's no standard** i.e. he can't stand up (*OED* 3). Dover Wilson suggested a pun on the Standard in Cornhill, a public water fountain (*OED* iii. 17), with associated *double-entendres* in run, go, lie (= urinate, excrete), ll. 17–18.
17 **run** i.e. from the enemy
18 **go** walk. Tilley cites 'He may ill run that cannot go.' (R208).
lie (a) lie down; (b) tell lies. 'To lie like a dog' was proverbial (Dent D510.2).
24 **in case** prepared (i.e. drunk, and hence valiant)
31 **natural** idiot; the point of the quibble is that a monster is by definition unnatural.

STEPHANO Trinculo, keep a good tongue in your head. If you
prove a mutineer, the next tree! The poor monster's my
subject, and he shall not suffer indignity.
CALIBAN I thank my noble lord. Wilt thou be pleased to
hearken once again to the suit I made to thee?
STEPHANO Marry, will I. Kneel and repeat it. I will stand,
and so shall Trinculo.

> *Enter Ariel, invisible*

CALIBAN As I told thee before, I am subject to a tyrant, a   40
sorcerer that by his cunning hath cheated me of the is-
land.
ARIEL Thou liest.
CALIBAN (*to Trinculo*) Thou liest, thou jesting monkey, thou!
I would my valiant master would destroy thee! I do not
lie.
STEPHANO Trinculo, if you trouble him any more in's tale,
by this hand, I will supplant some of your teeth.
TRINCULO Why, I said nothing.
STEPHANO Mum, then, and no more.—Proceed.   50
CALIBAN
  I say by sorcery he got this isle;
  From me he got it. If thy greatness will
  Revenge it on him—for I know thou dar'st,
  But this thing dare not—
STEPHANO That's most certain.
CALIBAN Thou shalt be lord of it, and I'll serve thee.
STEPHANO How now shall this be compassed? Canst thou
bring me to the party?

51–2 isle; | From me] THEOBALD; isle | From me, F

33 **keep . . . head** proverbial (Tilley T402)
34 **the next tree** i.e. to serve as a gallows
39.1 *Enter . . . invisible* Shakespeare had
used the same comic device for Puck, acting on Oberon's instructions to exacer-
bate the quarrel between Demetrius and
Lysander, *Dream* 3.2.354 ff. Similarly, the
invisible Faustus and Mephistopheles play
tricks on the Pope in Marlowe's *Doctor
Faustus* (1592), 7.50 ff. The 'robe for to
goo invisibell' listed in Henslowe's papers
as belonging to the Lord Admiral's Men
(eds. Foakes and Rickert, p. 325) suggests
that the device was not an uncommon

one on the Elizabethan stage.
40 ff Caliban's prose begins to take on the
rhythms of blank verse. F prints the scene
from Ariel's entrance on, including
Stephano's and Trinculo's speeches, as
verse.
48 **supplant** uproot (used of Antonio's
usurpation of Prospero at 2.1.269 and
3.3.70)
49–50 **nothing./Mum** proverbial: 'I will
say nothing but mum' (Tilley N279)
56 The pentameter rhythm becomes clear if
thee is emphasized: Caliban says, 'I'll
serve thee, not Prospero.'

CALIBAN

　Yea, yea, my lord. I'll yield him thee asleep,

　Where thou mayst knock a nail into his head.   60

ARIEL Thou liest, thou canst not.

CALIBAN

　What a pied ninny's this! Thou scurvy patch!

　I do beseech thy greatness, give him blows,

　And take his bottle from him. When that's gone,

　He shall drink naught but brine, for I'll not show him

　Where the quick freshes are.

STEPHANO Trinculo, run into no further danger. Interrupt

　the monster one word further, and by this hand, I'll turn

　my mercy out o' doors and make a stockfish of thee.

TRINCULO Why, what did I? I did nothing! I'll go farther off.   70

STEPHANO Didst thou not say he lied?

ARIEL Thou liest.

STEPHANO Do I so? Take thou that! (*Beats Trinculo*) As you

　like this, give me the lie another time!

TRINCULO I did not give the lie! Out o' your wits and hearing

　too? A pox o' your bottle! This can sack and drinking do.

　A murrain on your monster, and the devil take your

　fingers!

CALIBAN Ha, ha, ha!

STEPHANO Now forward with your tale. (*To Trinculo*)   80

　Prithee, stand further off.

CALIBAN

　Beat him enough. After a little time

　I'll beat him too.

STEPHANO Stand farther.—Come, proceed.

---

73 *Beats Trinculo*] ROWE (*Beats him*); *not in* F　　80 *To Trinculo*] This edition; *not in* F

60 **knock . . . head** on the biblical model of
Jael's murder of the sleeping Sisera
(Judges 4 : 21)

62 **pied** Caliban alludes to Trinculo's particoloured jester's costume.
patch jester, fool. The word is probably
cognate with Italian *pazzo* (fool), but
Shakespeare seems to have associated it
with the jester's particoloured garments:
compare 'a patched fool', *Dream* 4.1.215.

66 **quick freshes** the 'fresh springs' of

1.2.338
69 **stockfish** dried and salted cod, beaten to
tenderize it before cooking. 'To beat like a
stockfish' was proverbial (Tilley S867).

74 **give me the lie** call me a liar

77 **murrain** plague

81 **stand further off** i.e. further than Trinculo himself has offered to stand, l. 70.
Furness thought that Stephano addresses
this to Caliban, finding his 'ancient and
fish-like smell' offensive.

CALIBAN

> Why, as I told thee, 'tis a custom with him
> I' th' afternoon to sleep. There thou mayst brain him,
> Having first seized his books; or with a log
> Batter his skull, or paunch him with a stake,
> Or cut his weasand with thy knife. Remember
> First to possess his books; for without them   90
> He's but a sot, as I am, nor hath not
> One spirit to command—they all do hate him
> As rootedly as I. Burn but his books.
> He has brave utensils, for so he calls them,
> Which when he has a house, he'll deck withal.
> And that most deeply to consider is
> The beauty of his daughter. He himself
> Calls her a nonpareil. I never saw a woman
> But only Sycorax, my dam, and she;
> But she as far surpasseth Sycorax   100
> As great'st does least.

STEPHANO   Is it so brave a lass?

CALIBAN

> Ay, lord, she will become thy bed, I warrant,
> And bring thee forth brave brood.

STEPHANO Monster, I will kill this man. His daughter and I
will be king and queen—save our graces!—and Trinculo and thyself shall be viceroys. Dost thou like the plot,
Trinculo?

TRINCULO Excellent.

STEPHANO Give me thy hand. I am sorry I beat thee. But
while thou liv'st keep a good tongue in thy head.   110

CALIBAN

> Within this half hour will he be asleep.
> Wilt thou destroy him then?

85–6 **'tis a custom . . . sleep** Old Hamlet's custom too, providing Claudius with a similar opportunity for murder (*Hamlet* 1.5.59–60)
86 **There** then (Abbott 70)
88 **paunch him** stab him in the belly
91 **sot** fool
94 **utensils** either magical paraphernalia or simply household goods: l. 95 would admit of either meaning. The word was accented on the first syllable until the mid-eighteenth century.
99 **she** for 'her': Abbott 211
101 **brave** good looking
106 **plot** Kermode suggests an overtone of the additional Elizabethan sense 'the outline of a play or entertainment' (*OED* ii 45).
110 **keep . . . head** Stephano repeats his proverbial warning of l. 33.

22. The clown Richard Tarlton
(d. 1588) playing the pipe
and tabor, wash drawing.

STEPHANO   Ay, on mine honour.

ARIEL This will I tell my master.

CALIBAN

Thou mak'st me merry. I am full of pleasure;
Let us be jocund. Will you troll the catch
You taught me but whilere?

STEPHANO

At thy request, monster, I will do reason, any reason.
Come on, Trinculo, let us sing.

   They sing

         Flout 'em and cout 'em
         And scout 'em and flout 'em,   120
            Thought is free.

118.1 *They sing*] F (*Sings*)

115 **troll the catch** sing the round
116 **but whilere** a little while ago
117 **reason, any reason** anything reason-
     able
119 **cout** The word must, like *flout* and *scout*,
     imply ridicule. It is generally emended,
     following Rowe, to 'scout', to make the
     line accord with l. 120, but the similarity

between the two words exists only in a
modernized text: F reads 'cout' and
'skowt'. OED records *cout* only as a
variant of *colt* and *coot*, but it may be a
nonce word, or unrecorded slang.
120 **scout** mock, deride (*OED* v²)
121 proverbial (Tilley T244); quoted also by
     Maria in *Twelfth Night* 1.3.68

CALIBAN That's not the tune.
    Ariel plays the tune on a tabor and pipe
STEPHANO What is this same?
TRINCULO This is the tune of our catch, played by the picture
    of Nobody.
STEPHANO If thou beest a man, show thyself in thy likeness.
    If thou beest a devil, take't as thou list.
TRINCULO O, forgive me my sins!
STEPHANO He that dies pays all debts. I defy thee! Mercy
    upon us!   130
CALIBAN Art thou afeard?
STEPHANO No, monster, not I.
CALIBAN
    Be not afeard, the isle is full of noises,
    Sounds, and sweet airs, that give delight and hurt not.
    Sometimes a thousand twangling instruments
    Will hum about mine ears; and sometime voices,
    That if I then had waked after long sleep,
    Will make me sleep again, and then in dreaming
    The clouds methought would open and show riches
    Ready to drop upon me, that when I waked   140
    I cried to dream again.
STEPHANO This will prove a brave kingdom to me, where I
    shall have my music for nothing.

122.1 *tabor and pipe* A tabor is a drum that
hangs at one's side, and the pipe is a
tabor-pipe, designed to be played with one
hand. The combination was associated
with rustic dances and popular merry-
making. See Fig. 22.

124–5 **the picture of Nobody** This would, of
course, be invisible. The personification of
Nobody has a long history, beginning
with the Cyclops episode in the *Odyssey*.
Furness noted an anonymous comedy
called *No-body and Some-body* (*c.* 1606)
with a picture of a man with head and
limbs but no body on the title page; the
play's publisher, John Trundle, also used
'the sign of No-body' as his shop sign.
Presumably this or some other topical
allusion is involved.

126–7 **If . . . list** Kermode compares the expression 'The devil take it'. 'Take it as you
list' is proverbial (Tilley T27). Kittredge

suggested that Stephano in his drunkenness has mixed his terms up: it is the devil
who should be asked to appear in his true
shape, the man who should be invited to
take Stephano's challenge any way he
pleases.

129 **He . . . debts** proverbial (Tilley D148)

129–30 **Mercy upon us** Stephano's bravado
collapses.

135 **twangling** 'Describing a resonant
sound of the nature of a twang, but
thinner and continuous or repeated'
(*OED*). The word is usually used
pejoratively.

143 **music for nothing** There is probably a
topical overtone to this: payments to
court musicians constituted an increasingly large item in the royal budget,
particularly for masques and the grand
balls they included, and the royal family's
extravagance in the provision of such

CALIBAN When Prospero is destroyed.

STEPHANO That shall be by and by. I remember the story.

TRINCULO The sound is going away. Let's follow it, and after
do our work.

STEPHANO Lead, monster, we'll follow. I would I could see
this taborer; he lays it on.

TRINCULO (*to Caliban*) Wilt come? I'll follow Stephano.   150

                                                    Exeunt

3.3   *Enter Alonso, Sebastian, Antonio, Gonzalo, Adrian,
       Francisco*

GONZALO (*to Alonso*)
  By' r lakin, I can go no further, sir,
  My old bones aches. Here's a maze trod indeed
  Through forth-rights and meanders! By your patience,
  I needs must rest me.

ALONSO   Old lord, I cannot blame thee,
  Who am myself attached with weariness
  To th' dulling of my spirits. Sit down and rest.
  Even here I will put off my hope, and keep it
  No longer for my flatterer. He is drowned
  Whom thus we stray to find, and the sea mocks
  Our frustrate search on land. Well, let him go.   10

ANTONIO (*aside to Sebastian*)
  I am right glad that he's so out of hope.
  Do not for one repulse forgo the purpose
  That you resolved t' effect.

150 *to Caliban*] This edition; *not in* F
3.3.0.2 *Francisco*] F *adds '&c.'* (*but compare 5.1.57.2 ff.*)

entertainments was a continual subject of
anti-court polemic. The King made a
number of attempts around this period to
limit such expenditures. The matter is
wittily explored in Ben Jonson's masque
Love Restored (1612), one of the least ex-
pensive of the Jacobean court productions.
149 **lays it on** i.e. bangs his drum
150 Steevens, followed by most editors including Kermode and Barton, inserts a
comma before 'Stephano'. But F 's punctuation needs no emendation: *Wilt come?*

is addressed to Caliban, who at Trinculo's
invitation exits first, followed in turn by
Stephano and Trinculo.
3.3.1 **By'r lakin** by our ladykin, a mild form
of 'by our Lady'
3 **forth-rights and meanders** paths that are
sometimes straight and sometimes winding
5 **attached** seized; a legal metaphor
6 **To . . . spirits** to the point at which my
spirits are dulled
10 **frustrate** vain

SEBASTIAN (*aside to Antonio*) The next advantage
  Will we take throughly.
ANTONIO (*aside to Sebastian*) Let it be tonight;
  For now they are oppressed with travail, they
  Will not nor cannot use such vigilance
  As when they are fresh.
SEBASTIAN (*aside to Antonio*) I say tonight. No more.
      Solemn and strange music, and Prospero on the top,
      invisible
ALONSO
  What harmony is this? My good friends, hark!
GONZALO Marvellous sweet music!
      Enter several strange shapes bringing in a banquet, and
      dance about it with gentle actions of salutations; and
      inviting the King, etc., to eat, they depart
ALONSO
  Give us kind keepers, heavens! What were these?   20
SEBASTIAN
  A living drollery! Now I will believe
  That there are unicorns; that in Arabia
  There is one tree, the phoenix' throne, one phoenix
  At this hour reigning there.

17.1 *Prospero*] F (*Prosper*)   19.1 ff. *As in* WILSON; F *places this and the stage direction at* 17.1
together after 'fresh.', *l. 17*

14 **throughly** thoroughly
15 **now** now that
17.1 **the top** a technical term for the level above the upper stage gallery, within which the musicians sat. Prospero oversees the action from the highest point possible on the Jacobean stage. J. C. Adams provides an elaborate and highly speculative reconstruction of the original staging of the scene from this point onwards in *The Globe Playhouse* (New York, 1954), pp. 319–22.
20 **keepers** guardian angels
21 **A living drollery** either a comic play in real life, or a living caricature. The relevant senses of *drollery* are 'comic entertainment', 'comic picture or caricature', and 'puppet show'. Since Steevens's time the phrase has almost in-variably been glossed 'a living puppet show', but either of the other meanings seems preferable. Shakespeare uses *drollery* in *2 Henry IV* 2.1.156 to mean a comic picture to hang on the wall, and Jonson in the Induction to *Bartholomew Fair* (1614) uses the word to refer to plays like *The Tempest*: 'those that beget tales, tempests, and such-like drolleries . . .'(ll. 129–30).
23 **phoenix** a mythical Arabian bird, said to exist only one at a time, to nest in a single tree, and to reproduce by expiring in flame and then resurrecting itself from its own ashes. Shakespeare made it the subject of an elegant and elusive allegorical poem, *The Phoenix and the Turtle*, published in 1601.

ANTONIO   I'll believe both;
And what does else want credit, come to me,
And I'll be sworn 'tis true. Travellers ne'er did lie,
Though fools at home condemn 'em.

GONZALO   If in Naples
I should report this now, would they believe me?
If I should say I saw such islanders—
For certes these are people of the island—   30
Who though they are of monstrous shape, yet note
Their manners are more gentle-kind than of
Our human generation you shall find
Many, nay almost any.

PROSPERO (*aside*)   Honest lord,
Thou hast said well; for some of you there present
Are worse than devils.

ALONSO   I cannot too much muse
Such shapes, such gesture, and such sound expressing,
Although they want the use of tongue, a kind
Of excellent dumb discourse.

PROSPERO (*aside*)   Praise in departing.

FRANCISCO
They vanished strangely.

SEBASTIAN   No matter, since   40
They have left their viands behind; for we have
     stomachs.
Will't please you taste of what is here?

ALONSO   Not I.

---

29 islanders] F2; islands F1   32 gentle-kind] THEOBALD; gentle, kind F (*see note*)   36 muse]
F1; muse, F4

---

25 **what . . . credit** anything else unbeliev-
    able
26 **Travellers ne'er did lie** Tilley records 'A
    traveller may lie with authority' (T476)
30 **certes** certainly; by Shakespeare's time
    the word was archaic and exclusively
    literary.
32 **gentle-kind** either having the gracious-
    ness of nobility or noble-mannered. F's
    'gentle, kind' is probably a compound ad-
    jective; such pointing is not uncommon

in the Folio: compare 'marriage-blessing'
(4.1.106) and 'thousand-thousand'
(3.1.91).
36 **muse** marvel at. Shakespeare does not
    elsewhere use the word transitively, but
    OED (s.v. 3b and 10) cites several
    analogous examples in the period.
39 **Praise in departing** proverbial, meaning
    'keep your praise till the end' (Tilley P83)
41 **viands** food
    stomachs good appetites

GONZALO

Faith, sir, you need not fear. When we were boys,
Who would believe that there were mountaineers
Dewlapped like bulls, whose throats had hanging at
    'em
Wallets of flesh?—or that there were such men
Whose heads stood in their breasts?—which now
    we find
Each putter-out of five for one will bring us
Good warrant of.

ALONSO   I will stand to, and feed,
Although my last—no matter, since I feel   50
The best is past. Brother, my lord the Duke,
Stand to and do as we.
        Thunder and lightning.
        Enter Ariel, like a harpy, claps his wings upon the
        table, and with a quaint device the banquet vanishes

ARIEL

You are three men of sin, whom Destiny,

52 to] F4 (*after* too F)

44 **mountaineers** mountain-dwellers
46 **Wallets** wattles
46–7 **men | Whose heads stood in their breasts** alluded to also in *Othello* 1.3.144–5: 'The anthropophagi, and men whose heads | Do grow beneath their shoulders.'
48 **putter-out of five for one** London brokers provided a form of insurance to allow travellers to recoup their expenses: the traveller deposited a sum of money with the broker before departing, which was repaid fivefold if he returned with proof that he had reached his destination, but failing that was forfeited to the broker. The 'putter-out', then, is either the traveller, who invests his money at the rate of five to one, or the broker, who pays at that rate and reports the traveller's tales.
49 **stand to** 'To set to work, fall to; *esp.* to begin eating' (*OED* s. stand 101b)
52.2 *Ariel . . . harpy* The episode is based on *Aeneid* iii. 225 ff.: Aeneas and his companions take shelter on the Strophades, the islands where the harpies live. The Trojans prepare a feast; but, as they are about to eat, the dreadful creatures swoop down on them, befouling and devouring their food. The sailors attempt to drive the harpies off, but find them invulnerable, and their leader, the witch Celaeno, sends Aeneas away with a dire prophecy. Harpies had the faces and breasts of young women, the wings and bodies of birds, and talons for hands. *Enter . . . like a harpy* may imply that Ariel enters flying (see the Introduction, p. 2).
52.3 *with a quaint device* by means of an ingenious mechanism (which probably whisked the banquet down through a hole in the table, perhaps with the assistance of a stagehand hidden beneath it). The admiring vagueness of the wording suggests another of Crane's revisions. See the Introduction, pp. 57–8. For an elaborate but highly conjectural attempt to reconstruct the mechanics of this scene, see J. C. Adams, 'The Staging of *The Tempest*, 3.3.', *RES* 14 (1938), pp. 404–19.
53–82 For all its recollections of the *Aeneid*, the speech's syntax and tone are Prospero's, and he takes credit for its substance at ll. 85–6.

That hath to instrument this lower world
And what is in't, the never-surfeited sea
Hath caused to belch up you, and on this island,
Where man doth not inhabit—you 'mongst men
Being most unfit to live. I have made you mad;
And even with such-like valour men hang and drown
Their proper selves.

      Alonso, Sebastian, etc. draw their swords

               You fools! I and my fellows   60
Are ministers of Fate—the elements
Of whom your swords are tempered may as well
Wound the loud winds, or with bemocked-at stabs
Kill the still-closing waters, as diminish
One dowl that's in my plume. My fellow ministers
Are like invulnerable. If you could hurt,
Your swords are now too massy for your strengths,
And will not be uplifted. But remember—
For that's my business to you—that you three
From Milan did supplant good Prospero,   70
Exposed unto the sea, which hath requit it,
Him and his innocent child; for which foul deed,
The powers delaying, not forgetting, have
Incensed the seas and shores, yea all the creatures
Against your peace. Thee of thy son, Alonso,

---

60 *Alonso . . . swords*] CAMBRIDGE; *They draw their swords* HANMER; *not in* F   65 *plume*] ROWE; *plumbe* F

---

54 **to instrument** as its instrument
59 **such-like valour** insane courage, as opposed to true heroism
60 **Their proper selves** an intensive form of 'themselves': Abbott 16
61–4 **elements . . . waters** The swords are forged by the action of fire on metal refined from earth; these elements are ineffective against the other two elements, the air of winds and 'the still-closing waters'.
62 **whom** For 'which' (see Abbott 264); but Kermode suggests that F's comma may imply that Shakespeare 'changed direction here, after beginning to say "We are the elements of which your swords are

tempered; therefore they cannot hurt us."'
    tempered both compounded and hardened
64 **still-closing** that close as soon as they are parted
65 **dowl** the filament of a feather; the smallest feather
66 **like** alike
67 **massy** heavy
71 **Exposed** The object of the verb is 'Him and his innocent child'.
    hath requit it has now avenged the deed
73 **delaying, not forgetting** Compare the proverb 'God stays long but strikes at last' (Tilley G224).

They have bereft; and do pronounce by me
Ling'ring perdition, worse than any death
Can be at once, shall step by step attend
You and your ways; whose wraths to guard you from,
Which here, in this most desolate isle, else falls   80
Upon your heads, is nothing but heart's sorrow,
And a clear life ensuing.
> *He vanishes in thunder. Then, to soft music, enter the*
> *shapes again, and dance with mocks and mows, and*
> *carrying out the table* ⌈*they depart*⌉

PROSPERO
Bravely the figure of this harpy hast thou
Performed, my Ariel; a grace it had, devouring.
Of my instruction hast thou nothing bated
In what thou hadst to say; so with good life
And observation strange my meaner ministers
Their several kinds have done. My high charms work,
And these, mine enemies, are all knit up
In their distractions. They now are in my power;   90
And in these fits I leave them, while I visit
Young Ferdinand, whom they suppose is drowned,

82.3 *they depart*] This edition; *not in* F

77 **Ling'ring perdition** slow and continuous
destruction; a protracted hell-on-earth.
The threat has been exemplified in the
emblem of the disappearing banquet, tan-
talizing, but promising only slow starva-
tion. (Prospero uses the obsolete, poetic
word 'perdition' in another sense at
1.2.30.) Syntactically, the phrase is the
centre of an *apo koinu* construction, a
grammatical zeugma, being the object of
pronounce and the subject of *shall ... attend*. The figure is a common one in
Shakespeare (compare 5.1.59–60.), and
in English until the mid-seventeenth century, though editors since the eighteenth
century have tended to punctuate or
emend it out of existence wherever possible. Abbott (224) gives many examples,
though he treats the figure, erroneously,
as an omitted relative.
79 **whose wraths** i.e. those of the powers of
l. 73
81 **is nothing** there is no alternative

82 **clear** blameless
82.2 *mocks and mows* 'To mock and mow'
(mouth derisively) was proverbial: see
Tilley M1030.
84 **devouring** Ariel's removal of the banquet is characterized with a classical
reminiscence: in the *Aeneid* the harpies
actually devour the Trojans' meal. Kermode remarks that 'the word is specially
appropriate to a Harpy, and Ariel has just
removed the food, perhaps in such a way
as to suggest that he had devoured it'.
85 **bated** omitted
86 **so** in the same way
   with good life Prospero praises the spirits
   for both vitality and naturalness in their
   performance.
87 **observation strange** wonderful attentive-
   ness (to Prospero's commands)
88 **Their several kinds have done** have performed their various roles
92 **whom** for 'who': Abbott treats this
   as a confusion of two constructions,

And his and mine loved darling.   *Exit above*

GONZALO
I' th' name of something holy, sir, why stand you
In this strange stare?

ALONSO   O, it is monstrous, monstrous!
Methought the billows spoke and told me of it,
The winds did sing it to me; and the thunder,
That deep and dreadful organ-pipe, pronounced
The name of Prosper: it did bass my trespass.
Therefore my son i' th' ooze is bedded; and   100
I'll seek him deeper than e'er plummet sounded,
And with him there lie mudded.   *Exit*

SEBASTIAN   But one fiend at a time,
I'll fight their legions o'er.

ANTONIO   I'll be thy second.
                              Exeunt Sebastian and Antonio

GONZALO
All three of them are desperate: their great guilt,
Like poison given to work a great time after,
Now 'gins to bite the spirits. I do beseech you
That are of suppler joints, follow them swiftly,
And hinder them from what this ecstasy
May now provoke them to.

ADRIAN   Follow, I pray you.
                              All exeunt

93 *Exit above*] CAMBRIDGE; *Exit Prospero from above* THEOBALD; *not in* F   103.1 *Exeunt . . .*
Antonio] MALONE.; *Exeunt.* F   109.1 *All exeunt*] F (*Exeunt omnes.*)

Ferdinand who, they suppose, is drowned,' and 'whom they suppose to be drowned' (410).

93 **mine** 'Mine, hers, theirs are used as pronominal adjectives *before* their nouns. That *mine* should be thus used is not remarkable, as in E[arly] E[nglish] it was interchangeable with *my*, and is often used by Shakespeare where we should use *my*' (Abbott 238). The use of the form before a consonant, while unusual, is not unique in Shakespeare: compare *Antony* 4.8.18: 'Mine nightingale, we have beat them . . .'

95 **stare** 'A condition of amazement, horror, admiration, etc., indicated by staring' (*OED*, *sb.*² 2)

99 **bass my trespass** (a) sing out my guilt in a bass voice: in the harmonic image, the thunder takes the bass line while the winds of l. 97 sing the higher parts; (b) provide a (musical) ground or basis for the revelation of my guilt; (c) with a play on 'utter the baseness of my guilt'. Alonso hears Ariel's speech as the assertion of retributive justice through the harmony of nature.

100 **Therefore** for that

103 **o'er** one after another

104 **desperate** (a) in despair; (b) dangerously reckless

106 **spirits** vital powers

108 **ecstasy** madness

4.1   *Enter Prospero, Ferdinand, and Miranda*
PROSPERO (*to Ferdinand*)
 If I have too austerely punished you
 Your compensation makes amends, for I
 Have given you here a third of mine own life,
 Or that for which I live; who once again
 I tender to thy hand. All thy vexations
 Were but my trials of thy love, and thou
 Hast strangely stood the test. Here, afore heaven,
 I ratify this my rich gift. O Ferdinand,
 Do not smile at me that I boast of her,
 For thou shalt find she will outstrip all praise,   10
 And make it halt behind her.
FERDINAND   I do believe it
 Against an oracle.

4.1.1 *to Ferdinand*] This edition; *not in* F 9 of her] This edition; her of F; her off F2 (*see note*)

4.1.1 **austerely** harshly, rigorously
 punished Prospero in l. 6 explains his
 actions as constituting 'trials of thy love',
 but punishment implies crimes to be
 atoned for, and the choice of words here
 recalls his baseless charges against Ferdinand at 1.2.454 ff. See the Introduction, p. 29.
3–4 **a third . . . live** The simplest explanation
 is that a third merely signifies a very important part, as it clearly does in Prospero's
 later declaration that 'Every third thought
 shall be my grave' (5.1.311). Alternatively, we may take it to mean that for a third
 of his life Miranda has been the centre of
 his existence (this would put Prospero's
 age at 45). Editors who wish to extract
 more biographical significance than this
 from the statement are faced with finding
 the other two-thirds, a question on which
 the play gives us little help. Editorial
 debate over the passage began with Theobald, who emended 'third' to 'thread'.
 Capell thought no emendation necessary,
 and that the three thirds were 'his realm,
 his daughter, himself'; Kermode finds this
 plausible. Other editors have suggested
 that Prospero was thinking of his dead
 wife as one of the thirds: if so, this would
 be only the second time she has entered
 his mind in the course of the play.

4 **who** for 'whom', Abbott 274
7 **strangely** wonderfully; compare 'by observation strange', 3.3.87.
9 **of her** It seems best to treat F's 'her of' as
 a compositor's inversion—the syntax,
 while not impossible, is without parallel
 in Shakespeare. A similar inversion
 occurs at 4.1.193, 'Come, hang on them
 this line', where 'on them' has been
 regularly emended to 'them on' since
 Rowe. F2 changed 'boast her of' to 'boast
 her off'; this may be considered merely a
 spelling variant, and has been accepted by
 most editors: Kermode explains that
 'boast her off' means 'cry up her praises'.
 But, as Dover Wilson observed, there are
 no Shakespearian parallels for this usage
 either, so F2's version is open to the same
 objection as the original reading. Moreover, the tone of 'boast her off' seems un-
 comfortably colloquial after the formality
 of the previous lines. Dover Wilson's
 emendation 'hereof', on the other hand,
 sounds rather too legalistic as a response
 to Ferdinand's smile.
11 **halt** limp
12 **Against an oracle** though an oracle
 should contradict it—an ambiguous assurance, if we think of Leontes (*Winter's
 Tale* 3.2)

PROSPERO
Then as my gift, and thine own acquisition
Worthily purchased, take my daughter. But
If thou dost break her virgin-knot before
All sanctimonious ceremonies may
With full and holy rite be ministered,
No sweet aspersion shall the heavens let fall
To make this contract grow; but barren hate,
Sour-eyed disdain, and discord shall bestrew   20
The union of your bed with weeds so loathly
That you shall hate it both. Therefore take heed,
As Hymen's lamps shall light you.
FERDINAND   As I hope
For quiet days, fair issue, and long life,
With such love as 'tis now, the murkiest den,
The most opportune place, the strong'st suggestion
Our worser genius can, shall never melt

---

13 gift] ROWE; guest F   17 rite] ROWE; right F

---

13 **gift** Rowe's emendation of F's 'guest'. T.
H. Howard-Hill notes that 'Crane's
preferred spelling was "guift"' (Crane,
p. 111). F in l. 8 above reads 'guift':
presumably the compositor's copy had
the same spelling here.
14 **purchased** won
15 **break . . . knot** Probably alluding to
Catullus lii. 27, 'Zonam soluere vir-
gineam', to untie the virgin's girdle; but
Prospero's *break* has implications that are
a good deal more violent than Catullus'
'untie', and in this context the *knot* may
also be a hard kernel, or the hard node of
a tree trunk: 'blunt wedges rive hard
knots' (*Troilus* 1.3.316). (For 'breaking' a
knot, see OED's citation from Moxon s.
knot 17.)
16 **sanctimonious** sacred
18 **aspersion** literally, sprinkling; the
heavenly rain that nourishes fruition.
Compare 3.1.75–6: 'Heavens rain grace
| On that which breeds between 'em!'.
21 **bed** with an overtone of 'seed-bed'
weeds The marriage bed was customarily
strewn with flowers.
loathly loathsome: 'rare in the seventeenth and eighteenth centuries' (*OED*).

Shakespeare also uses the word in *2 Henry
IV* and *Lear*.
23 **Hymen's lamps** Hymen was god of marriage, and his lamps are the wedding
torches. These were regarded as good
omens if they burned clear, bad if they
were smoky.
25 **den** originally the lair of a wild beast, and
by extension any enclosed hiding place,
generally with dangerous or unsavoury
overtones
26 **opportune** accented on the second syllable
suggestion temptation
27 **genius** 'The tutelary god or attendant
spirit allotted to every person at his birth,
to govern his fortunes and determine his
character, and finally to conduct him out
of the world', and from this, a person's
good and evil genius, 'the two mutually
opposed spirits (in Christian language
angels) by whom every person was supposed to be attended throughout his life'
(*OED*).
can can make
27–8 **melt . . . lust** The metaphor is continued in l. 55–6: 'The white cold virgin
snow upon my heart | Abates the ardour
of my liver.'

Mine honour into lust, to take away
The edge of that day's celebration
When I shall think or Phoebus' steeds are foundered,   30
Or night kept chained below.

PROSPERO   Fairly spoke.
Sit then and talk with her, she is thine own.
What, Ariel! My industrious servant Ariel!
  Enter Ariel

ARIEL
What would my potent master? Here I am.

PROSPERO
Thou and thy meaner fellows your last service
Did worthily perform, and I must use you
In such another trick. Go, bring the rabble
O'er whom I give thee pow'r here to this place.
Incite them to quick motion, for I must
Bestow upon the eyes of this young couple   40
Some vanity of mine art: it is my promise,
And they expect it from me.

ARIEL   Presently?

PROSPERO   Ay, with a twink.

ARIEL   Before you can say 'come' and 'go',
               And breathe twice, and cry 'so, so',
               Each one, tripping on his toe,
               Will be here with mop and mow.
               Do you love me, master? No?

PROSPERO
Dearly, my delicate Ariel. Do not approach

28 **to** so as to
29 **edge** ardour, with explicit sexual con-
    notations, as in *Hamlet* 3.2.249: 'It
    would cost you a groaning to take off
    mine edge.'
29–30 **that day's celebration | When** the
    celebration of that day on which
30–1 **or . . . below** either that the sun's
    horses have collapsed through overriding
    (so that the wedding night will never
    come) or that night is forcibly prevented
    from arriving
33 **What** now then
35 **meaner fellows** the 'meaner ministers' of
    3.3.87
37 **trick** ingenious artifice; here, with *per-*

form (l. 36), including specifically theatrical implications
  rabble the 'meaner fellows'; the word is
  invariably pejorative
39 **motion** 'Perhaps with a trace of the sense
    "puppet show"' (Kermode)
41 **vanity** a trifling display, as opposed to the
    raising of the storm and Ariel's porten-
    tous performance on the beach
42 **Presently** immediately
43 **with a twink** in the twinkling of an eye
    (literally, the time it takes to wink)
47 **mop and mow** Both words mean
    grimace. The rabble's previous perfor-
    mance included '*mocks and mows*': see
    3.3.82.2, and note.

Till thou dost hear me call.

ARIEL   Well, I conceive.   *Exit*   50

PROSPERO (*to Ferdinand*)

    Look thou be true; do not give dalliance

    Too much the rein. The strongest oaths are straw

    To th' fire i' th' blood. Be more abstemious,

    Or else good night your vow.

FERDINAND   I warrant you, sir,

    The white cold virgin snow upon my heart

    Abates the ardour of my liver.

PROSPERO   Well.

    Now come, my Ariel. Bring a corollary,

    Rather than want a spirit. Appear, and pertly!

      Soft music

    No tongue! All eyes! Be silent!

      Enter Iris

51 *to Ferdinand*] This edition; *not in* F

50 **conceive** understand

51 **true** faithful to your word: Prospero returns to the question of Ferdinand's chastity.
dalliance originally simply conversation; by Chaucer's time the primary sense was 'flirtation, amorous toying'.

51–2 **give . . . rein** 'To give one the reins' was proverbial (Tilley B671).

53 **Be more abstemious** This implies some kind of intense commerce between Ferdinand and Miranda. Most editors assume that when Prospero turns his attention to the lovers, they are embracing; but given the hyperbolic nature of Prospero's fears, they may merely be holding hands, deep in conversation, or looking adoringly into each other's eyes. How a director chooses to stage the moment will depend on the extent to which he sees Prospero's libidinous apprehensions as justified.

54 **warrant** Perhaps monosyllabic, as the word frequently is in the period.

55 **The . . . heart** either the thought of Miranda enshrined in his heart, or his own chaste love for her. 'As chaste as ice' was proverbial (Tilley I1).

56 **liver** in the old physiology, the seat of physical love and violent passion

57 **a corollary** one too many, 'a surplussage' (*OED*, citing Cotgrave)

58 **want** lack
pertly smartly, briskly (*OED* 3)

59 **No tongue** 'Those who are present at incantations are obliged to be strictly silent, "else," as we are afterwards told, "the spell is marr'd"' (Johnson).
All eyes Despite their speeches and songs, masques are primarily spectacles, and Prospero's demand for the attention of his audience emphasizes the crucial importance of the visual sense in these courtly entertainments. For a discussion of the masque, see the Introduction, pp. 43–50.

59.1 *Enter Iris* Iris is the messenger of the gods, and goddess of the rainbow. The masque she introduces is a betrothal masque, a concept that is apparently Shakespeare's invention. *As You Like It* includes a brief wedding masque presided over by Hymen (5.4.108 ff.), and the Jacobean court had recently seen two masques by Ben Jonson and Inigo Jones celebrating important aristocratic marriages: *Hymenaei* in 1606, for the wedding of the Earl of Essex and Lady Frances Howard, and *The Haddington Masque* in 1608, for one of James's Scottish favourites, John Ramsey, Viscount Haddington. Campion's *Lord Hay's Masque* (1607), for the wedding of another of James's Scottish lords, may also have

IRIS

Ceres, most bounteous lady, thy rich leas   60
Of wheat, rye, barley, vetches, oats, and peas;
Thy turfy mountains, where live nibbling sheep,
And flat meads thatched with stover them to keep;
Thy banks with pionèd and twillèd brims,
Which spongy April at thy hest betrims
To make cold nymphs chaste crowns; and thy broom
    groves,
Whose shadow the dismissèd bachelor loves,
Being lass-lorn; thy poll-clipped vineyard,

68 poll-clipped] F (pole-clipt)

been in Shakespeare's mind. Iris, Ceres, and Juno (along with ten other goddesses and Somnus) had appeared in the first masque presented at the Jacobean court, Daniel's *Vision of the Twelve Goddesses* (1604). See the Introduction, pp. 43–6.

60 **Ceres** goddess of earth and patroness of agriculture
    leas meadows
61 **vetches** tares, grown for fodder
63 **stover** winter forage
64 **pionèd and twillèd** The phrase has prompted a great deal of debate and emendation. *OED* gives 'pion' meaning dig out or excavate, but cites no instance of its use as a verb other than in the form 'pioning' ('pioneers' were originally the foot soldiers who cleared the way and dug protective trenches in advance of the army). *Twilled* means woven or plaited in such a way as to produce a diagonal or twill pattern. Henley, followed by Kittredge, argued that the effects of erosion are being described; T. P. Harrison, developing a suggestion of Knight's, saw the pioning and twilling as man-made attempts to reinforce the land against erosion through channelling and woven supports (*MLN* 58 (1943), p. 422). If *pioned* is really a form of 'pion', either of these explanations would make good sense. A less persuasive critical tradition has taken the phrase to refer to flowers, reading *pioned* as 'peonied', of which it is an acceptable period spelling, and emending *twilled* to 'tuliped', 'lilied', etc. Without emendation, the phrase could be modernized to 'peonied and twilled'.

66 **broom groves** Editorial opinion since the mid-eighteenth century has found a problem here: broom, which is a shrub, not a tree, cannot be said to grow in groves. Hanmer emended to 'brown-groves', and this has received some support, notably from Kermode, who, however, stops short of adopting it. I prefer to treat the passage as evidence that broom *could* be said, however loosely or infrequently, to grow in groves—if this is a usage invented by Shakespeare, it is not the first one. W. P. Mustard (*MLN* 38 (1923), pp. 79–81) cited analogous examples in Milton and Pope of corn groves, and in Wordsworth of ivy groves; he also suggested that the broom was a Virgilian allusion: shepherds in the *Georgics* ii. 434 rest in the shade of *genista* (broom). The allusion is dubious, but the groves are indisputable. As for the association between broom and the lovelorn, Geoffrey Grigson notes that the plant's amorous and magical properties were highly regarded, and he cites several ballads in which broom figures significantly in magic spells designed to ensure the success of love affairs (*The Englishman's Flora* (1955), pp. 128–9).
67 **dismissèd bachelor** rejected suitor
68 **poll-clipped** pollarded or pruned: this seems a more likely reading than 'pole-clipped', hedged with poles.
    vineyard probably trisyllabic, as two of *OED*'s period spellings suggest (vynyearde, viniard), and the rhyme requires a stress on 'yard' here.

And thy sea-marge sterile and rocky-hard,
Where thou thyself dost air : the Queen o' th' sky,   70
Whose watery arch and messenger am I,
Bids thee leave these, and with her sovereign grace,
Here on this grass-plot, in this very place,
To come and sport. Her peacocks fly amain.

⌜*Juno's chariot appears suspended above the stage*⌝

Approach, rich Ceres, her to entertain.

Enter ⌜*Ariel as*⌝ *Ceres*

CERES

Hail, many-coloured messenger, that ne'er
Dost disobey the wife of Jupiter;
Who with thy saffron wings upon my flowers
Diffusest honey-drops, refreshing showers;
And with each end of thy blue bow dost crown   80
My bosky acres and my unshrubbed down,
Rich scarf to my proud earth : why hath thy queen
Summoned me hither to this short-grassed green?

---

74 Her] ROWE; here F   74.1 *Juno's . . . stage*] This edition; *Juno descends* F, *in the margin at*
l. 72   75.1 *Ariel as*] This edition; *not in* F

70 **Queen o' th' sky** Juno
71 **watery arch** Iris as the rainbow
72 **these** 'thy rich leas' and the other places just described
74 **peacocks** sacred to Juno, and here conceived as drawing her chariot
    amain at full speed
74.1 The stage direction in F reads '*Juno descends*'; editors have almost invariably assumed that it is misplaced, and belongs at line 102, when Ceres says 'Great Juno comes'. Only Barton in her New Penguin edition, and more recently Jowett in a detailed and wholly convincing article, have argued that it is correct as it stands. Jowett writes, 'the direction "*descend(s)*" does not necessarily, or even usually, indicate a descent to the stage. There was, on the contrary, what has been called "the convention of the floating deity", whereby the deity would be expected, upon appearing from the heavens, to remain suspended in the air rather than to come down to the stage. It is common enough to find such an appearance signalled by the word "*descend(s)*", and it would not normally be taken as indicat-

ing a descent to the stage unless this was specified' ('New Created Creatures: Ralph Crane and the Stage Directions in *The Tempest*', *Shakespeare Survey* 36 (Cambridge 1983), pp. 116–17). In subsequent stage directions relating to Juno's entrance, I am relying on Jowett's analysis of the scene.
75 **entertain** receive as a guest (*OED* 13)
75.1 *Ceres* For this as Ariel's role, see l. 167 and note.
78 **saffron wings** In *Aeneid* iv. 700 Iris comes 'croceis pinnis', on saffron-coloured wings.
81 **bosky** covered with bushes
    unshrubbed down bare plains
83 **short-grassed green** The masque takes place on a well-tended lawn, as opposed to the wild nature of the island. There may be a specific reference here to the green rushes that covered the floor of the stage in public theatres (as Bottom says, 'This green plot shall be our stage', *Dream* 3.1.3), or to the green cloth that carpeted the dancing area when the Banqueting House was set up for a masque. See the Introduction, pp. 2–3.

IRIS

A contract of true love to celebrate,
And some donation freely to estate
On the blessed lovers.

CERES   Tell me, heavenly bow,
If Venus or her son, as thou dost know,
Do now attend the Queen? Since they did plot
The means that dusky Dis my daughter got,
Her and her blind boy's scandalled company   90
I have forsworn.

IRIS   Of her society
Be not afraid. I met her deity
Cutting the clouds towards Paphos, and her son
Dove-drawn with her. Here thought they to have done
Some wanton charm upon this man and maid,
Whose vows are that no bed-right shall be paid
Till Hymen's torch be lighted; but in vain.
Mars's hot minion is returned again;
Her waspish-headed son has broke his arrows,

---

85 **estate** bestow
86 **bow** rainbow
87 **as** so far as
88–9 **Since . . . got** In Ovid's account, Venus and Cupid, in order to extend their dominion to the underworld, inspired in Pluto the love that prompted him to abduct Proserpine (*Metamorphoses* v. 359 ff.).
89 **dusky** both dark and melancholy. Ovid calls Pluto '*niger*' (black).
 Dis The name means 'wealth': Pluto as god of the underworld was also god of riches. Dis is the Latin translation of the Greek name, a shortened form of *Dives*, wealth.
90 **blind** Cupid was traditionally represented as blindfolded (hence 'love is blind').
 scandalled scandalous
92 **her deity** jocular usage, on the model of 'his worship', 'her majesty', etc.
93 **Paphos** in Cyprus, the centre of Venus' cult
94 **Dove-drawn** Doves were sacred to Venus, and drew her chariot. At the end of *Venus and Adonis*, the goddess

Yokes her silver doves, by whose swift aid
Their mistress mounted through the empty skies

In her light chariot quickly is conveyed,
Holding their course to Paphos, where their queen
Means to immure herself, and not be seen. (1190–94)

94–5 **to . . . charm** to inspire them with lust, as they did to Dis
96 **bed-right shall be paid** The bed-right involves the payment of a debt of homage to Hymen. Steevens, followed by many editors, modernized to 'bed-rite', but Sisson objects that 'a *right* is paid, not a *rite*, which is celebrated. And the marriage-bed confers *rights* not *rites* upon the newly-wedded' (*New Readings*, p. 50). 'Bed-rite' is certainly a misguided reading; but, as Kermode observes, 'the notions of tribute and ceremony were doubtless confused or amalgamated' (p. 170), and the aural pun is to the point.
97 **Till . . . lighted** till the marriage ceremony is performed
98 **Mars's hot minion** the lustful Venus, whose illicit passion for Mars was discovered and revealed by her husband Vulcan (*Metamorphoses* iv. 184 ff). *Minion* = darling, mistress (French, *mignon*).
99 **waspish-headed** irascible; and his arrows sting

Swears he will shoot no more, but play with sparrows, 100
And be a boy right out.

CERES   Highest Queen of state,
Great Juno comes; I know her by her gait.

⌜*Juno's chariot descends to the stage*⌝

JUNO

How does my bounteous sister? Go with me
To bless this twain, that they may prosperous be,
And honoured in their issue.

    ⌜*Ceres joins Juno in the chariot, which rises and*
    hovers above the stage.⌝ *They sing.*

JUNO

    Honour, riches, marriage-blessing,
    Long continuance, and increasing,
    Hourly joys be still upon you!
    Juno sings her blessings on you.

CERES

    Earth's increase, foison plenty, 110
    Barns and garners never empty,
    Vines with clust'ring bunches growing,
    Plants with goodly burden bowing;
    Spring come to you at the farthest,
    In the very end of harvest!
    Scarcity and want shall shun you;
    Ceres' blessing so is on you.

102.1 *Juno's . . . stage*] This edition; *not in* F   105.1–2 *Ceres . . . stage*] This edition; *not in* F
106 marriage-blessing] THEOBALD; marriage, blessing F (*see note to* 3.1.91)   110 CERES] THEO-
BALD; *no change of speaker in* F

---

100 **sparrows** emblematic of lechery and associated with Venus. Chaucer's Summoner
was 'as hoot . . . and lecherous as a sparwe'
(*Canterbury Tales*, Prologue 626); 'as lustful
as sparrows' was proverbial (Tilley 715).
101 **be . . . out** simply be a boy, give up his
status as the god of love
Queen of state majestic Queen, an epithet
for Juno (not a vocative)
102 **gait** bearing, carriage—not necessarily
implying that she walks on to the stage.
103 **Go with me** This is more likely to be an
invitation into the chariot than a proposal
to promenade about the stage.
110 **foison** abundance (used by Gonzalo at
2.1.161)

111 **garners** granaries
114–15 i.e., after autumn, may spring
return at once, and your years have no
winter. The benediction undoes the effects
of Ceres' allusion to the rape of Proserpine
(ll. 88–9), which is responsible for the fact
that winter exists in nature: Proserpine
spends six months on earth and six in the
underworld, and during the latter period
Ceres, in mourning for her daughter,
allows no crops to grow. For the conceit of
a winterless year, compare Spenser's Garden of Adonis ('There is continual spring
and harvest there | Continual, both meet-
ing at one time,' *Faerie Queene* III vi. 42).
Fanny Kemble cited Leviticus 26, 'And

FERDINAND

This is a most majestic vision, and
Harmonious charmingly. May I be bold
To think these spirits?

PROSPERO   Spirits, which by mine art   120
I have from their confines called to enact
My present fancies.

FERDINAND   Let me live here ever.
So rare a wondered father and a wife
Makes this place paradise.

    Juno and Ceres whisper, and send Iris on employment

PROSPERO   Sweet, now, silence!
Juno and Ceres whisper seriously.
There's something else to do. Hush, and be mute,
Or else our spell is marred.

IRIS

You nymphs called naiads of the windring brooks,

---

your threshing shall reach unto the vintage, and the vintage shall reach unto
sowing time' (noted in the Variorum),
and there are classical antecedents in Virgil's Eclogue 4 and in Book I of the
Metamorphoses.

119 **Harmonious charmingly** 'Charming'
has a double etymology, from Latin *car-
men*, a song, and Anglo-Saxon *cierm*, a
spell, reflected in the traditional association of music with magic.
May I be bold Would I be correct : for the
boldness involved, compare the expression 'to venture an opinion'.

121 **their confines** the natural elements which
they inhabit ; *confines* as a noun was normally accented on the first syllable, but the
metre here seems to require an iamb.

122 **fancies** the same word as 'fantasies',
and originally simply a shortened spelling. By Shakespeare's time 'fancy' had
overtones of the light, arbitrary, and
capricious : Prospero here recalls the tone
of his characterization of the masque as
'some vanity of mine art' (l. 41).

123 **wondered** wonderful, endowed with
wonders—possibly also alluding to
Miranda's name

123 **wife** This has been conclusively shown
by Jeanne Addison Roberts to be the
reading intended by the Folio's compositor. Early in the print run, the cross-
bar of the f broke off, transforming 'wife'
to 'wiſe'. Several copies of the Folio show
the letter in the process of breaking.
(*University of Virginia Studies in
Bibliography* 31 (1978), pp. 203 ff.)

124 **Sweet** The epithet could be used between men : 'Nay, take me with thee,
good sweet Exeter' (King Henry in
3 *Henry VI* 2.5.137). Editors who are disturbed about the propriety of Prospero
addressing Ferdinand as *sweet* either assume that Miranda is about to speak, and
direct the line to her (Dover Wilson,
Kittredge), or have the speech down to
'seriously' spoken by Miranda (Wright,
Elze).

128 **naiads** Shakespeare may have found a
recent precedent for naiads in masques in
Daniel's *Tethys' Festival*, performed in the
Banqueting House on 5 June 1610 as part
of the celebrations for the investiture of
Prince Henry as Prince of Wales.
windring the only citation for this word in
the *OED*, which considers it a misprint

With your sedged crowns, and ever harmless looks,
Leave your crisp channels, and on this green land   130
Answer your summons, Juno does command.
Come, temperate nymphs, and help to celebrate
A contract of true love. Be not too late.
    Enter certain nymphs
You sunburned sickle-men, of August weary,
Come hither from the furrow and be merry;
Make holiday; your rye-straw hats put on,
And these fresh nymphs encounter every one
In country footing.
    Enter certain reapers, properly habited. They join with
    the nymphs in a graceful dance, towards the end
    whereof Prospero starts suddenly and speaks, after
    which, to a strange hollow and confused noise, they
    heavily vanish
PROSPERO (*aside*)
I had forgot that foul conspiracy
Of the beast Caliban and his confederates   140
Against my life. The minute of their plot
Is almost come. (*To the spirits*)—Well done, avoid. No
  more.
    ⌈*Juno and Ceres ascend in their chariot*
    and the reapers exeunt⌉

142.1–2 *Juno . . . reapers*] This edition; *not in* F

for 'winding'. Some editors emend to 'wand'ring'. If it is a misprint, it is a peculiarly felicitous one; but it may also be a nonce word, and is perfectly comprehensible. Kermode compares 'twangling' (3.2.135), which, however, was in current usage.

129 **sedged crowns** garlands of sedge, a river plant
   harmless literally translating 'innocent'
130 **crisp** rippling (*OED* 2)
132 **temperate nymphs** Nymphs were associated with chaste Diana, and their invocation here continues the theme begun by the account of Venus' and Cupid's departure (91 ff.).
134 **of August weary** because this is the time of harvest

137 **encounter** join
138 **country footing** rustic dancing
138.1 **properly** appropriately
138.4 **strange . . . noise** The harmony of the dance music dissolves in discord. *Hollow* = not full-toned, sepulchral (*OED* 4). The phrase may involve an echo of the Strachey letter, where hovering birds are described as making 'a strange hollow and harsh howling' (see Appendix B, p. 215).
138.4–5 *they heavily vanish* at l. 142.
138.5 *heavily* reluctantly, sorrowfully
139–42 For a discussion of the importance of this moment, see the Introduction, p. 50.
142 **avoid** begone

FERDINAND

    This is strange. Your father's in some passion
    That works him strongly.

MIRANDA   Never till this day
    Saw I him touched with anger, so distempered.

PROSPERO (*to Ferdinand*)

    You do look, my son, in a moved sort,
    As if you were dismayed. Be cheerful, sir;
    Our revels now are ended. These our actors,
    As I foretold you, were all spirits, and
    Are melted into air, into thin air,   150
    And, like the baseless fabric of this vision,

---

146 *to Ferdinand*] This edition; *not in* F

144–5 Never . . . distempered Prospero's sudden rage has been anticipated earlier in the day in Act I, Scene 2, in his account of his brother's usurpation and in his behaviour to Ferdinand; his actions here are clearly related to the complex feelings expressed in both those instances.

145 distempered literally, having the temper, or proportion, of the bodily humours disturbed. The term implies a physiological basis for vexation.

146–7 You . . . dismayed Critics have been puzzled at Prospero's need to comfort Ferdinand, when it is Prospero who is distressed. But Ferdinand is witnessing a return of the irrational behaviour of 1.2.443 ff., when he was accused of treason, paralysed, threatened with imprisonment, and forced to carry logs; and his dismay at Prospero's disruption of what seemed a happy ending to his trials is dramatically perfectly coherent.

146 sort condition

148–58 Our revels . . . sleep Prospero's speech is based on a *topos* descending to Shakespeare from classical times. T. W. Baldwin cites a parallel in Palingenius, a passage adapted by Barnabe Googe; Kermode adds Chrysostom, and Job 20 : 6–8. The closest analogue was discovered by Steevens in a passage from *The Tragedy of Darius* by William Alexander, Earl of Stirling, published in 1603 and dedicated to King James:

    . . . let this worldly pomp our wits enchant,
    All fades, and scarcely leaves behind a
        token.
    Those golden palaces, those gorgeous
        halls,
    With furniture superfluously fair:
    Those stately courts, those sky-
        encount'ring walls
    Evanish all like vapours in the air.

(*Poetical Works*, eds. L. E. Kastner and H. B. Charlton (Edinburgh, 1921), i. 196, ll. 1557–62.) Prospero's sentiments should also be compared with Jonson's and Daniel's published statements on the ephemeral nature of the masque: see the prefatory remarks to Jonson's *Masque of Blackness* (1605) and *Hymenaei* (1606), and Daniel's *Vision of the Twelve Goddesses* (1605) and *Tethys' Festival* (1610). Daniel, for example, refers to masques as 'punctilios [i.e. trifles] of dreams and shows' (*Vision*, l. 268), and to their creators as 'the poor engineers for shadows, [who] frame only images of no result' (*Tethys*, ll. 45–7). By 1611, such characterizations had become almost a convention of the genre.

148 revels entertainment, with an overtone from the technical language of the masque, in which the term is used for the final dance between masquers and spectators, the physical assertion of social harmony and aristocratic community

149 foretold you previously told you (not 'predicted to you')

151 baseless fabric structure without foundation; both an edifice without a base and a purposeless contrivance

The cloud-capped towers, the gorgeous palaces,
The solemn temples, the great globe Itself,
Yea, all which it inherit, shall dissolve,
And, like this insubstantial pageant faded,
Leave not a rack behind. We are such stuff
As dreams are made on, and our little life
Is rounded with a sleep. Sir, I am vexed.
Bear with my weakness, my old brain is troubled.
Be not disturbed with my infirmity.   160
If you be pleased, retire into my cell,
And there repose. A turn or two I'll walk
To still my beating mind.

FERDINAND *and* MIRANDA   We wish your peace.   *Exeunt*

PROSPERO

Come with a thought!—I thank thee.—Ariel, come!

    Enter Ariel

---

163 *Exeunt*] THEOBALD; *Exit.* F   164 thee . . . come!] KERMODE (thee. Ariel: come.); thee Ariel. come F

152–3 **towers . . . itself** These are the realities of Prospero's comparison, but they are also the fabric of masque visions: *Hymenaei* (1606) takes place before an altar sacred to Juno; the setting for the main masque of *The Masque of Queens* (1609) is the House of Fame; for *Oberon* (1611) Inigo Jones constructed a splendid palace, turreted and domed; and both *Hymenaei* and *The Haddington Masque* (1608) included a great globe on a turning machine. And of course the *great globe itself* would have had a still more explicit theatrical resonance, one combining both world and stage, to an audience seeing the play at the Globe or the Blackfriars.

154 **which it inherit** who subsequently possess it; succeeding generations

155 **pageant** another technical term, like 'revels'. The basic meanings are 'scene acted upon the stage', 'stage on which scenes are performed', 'stage machine', 'tableau or allegorical device'; when extended into the moral sphere the word implied deception, trickery, specious or empty show.

156 **rack** cloud or mist driven by the wind (*OED* 3, 3b), perhaps also with a suggestion of the cloud effects of masque scenery

157 **on** of

158 **rounded** either surrounded (*OED* ii. 11)

—our little life then being a brief awaking from an eternal sleep—or rounded off, i.e. completed. *OED* (i. 4) opts for the latter, an uncommon usage for which this passage is the earliest citation by seventy-four years. Shakespearian parallels, however, argue for 'surrounded', in both literal and figurative contexts, e.g. 'the hollow crown | That rounds the mortal temples of a King' (*Richard II* 3.2.161); 'our exposure | How rank soever rounded in with danger' (*Troilus* 1.3.195–6). Kermode favours Wright's explanation that the word means 'crowned', and offers a number of parallels, including the example from *Richard II*; but in every one of his cases 'round' appears in conjunction with 'crown', or with words clearly implying it. Prospero is not conceiving of death as the reward or high point of life, but as its largest condition.

158–9 **I . . . troubled** Prospero's sudden sense of infirmity is discussed in the Introduction, p. 50.

160 **with** by

163 **beating** throbbing, agitated—compare 1.2.176, 'still 'tis beating in my mind'.

164 **with a thought** as soon as I think of you. 'As swift as thought' was proverbial (Tilley T240).

I thank thee addressed to the parting

ARIEL

Thy thoughts I cleave to. What's thy pleasure?

PROSPERO

Spirit, we must prepare to meet with Caliban.

ARIEL

Ay, my commander. When I presented Ceres
I thought to have told thee of it, but I feared
Lest I might anger thee.

PROSPERO

Say again, where didst thou leave these varlets?   170

ARIEL

I told you, sir, they were red-hot with drinking,
So full of valour that they smote the air
For breathing in their faces, beat the ground
For kissing of their feet; yet always bending
Towards their project. Then I beat my tabor,
At which like unbacked colts they pricked their ears,
Advanced their eyelids, lifted up their noses
As they smelt music. So I charmed their ears
That calf-like they my lowing followed through
Toothed briars, sharp furzes, pricking gorse, and thorns,   180
Which entered their frail shins. At last I left them
I' th' filthy-mantled pool beyond your cell,

---

182 filthy-mantled] CAMBRIDGE; filthy mantled F

Ferdinand and Miranda. Some editors emend *thee* to 'you' or 'ye', but there is no problem about Prospero addressing to one of the lovers the thanks intended for both of them.

167 **When I presented Ceres** most likely, 'when I played the role of Ceres in the masque'; alternatively, 'when I produced the masque of Ceres'. Kermode adds a third, less likely possibility: 'as "presenter" of the masque, introduced her [Ceres] while playing the part of Iris.' But sheer theatrical economy would argue in favour of Shakespeare using his singer Ariel in a major role in the masque of Prospero's spirits, and Iris is not required to sing.

170 **varlets** clearly pejorative here, though the original meaning, servants or

menials, was still current in Shakespeare's time.

174 **bending** aiming

175 **tabor** Ariel's side-drum (see 3.2.122.1)

176 **unbacked** never ridden, and hence unbroken

177 **Advanced** raised; compare 1.2.409, 'The fringèd curtains of thine eye advance'.

178 **As** as if

180 **furzes, gorse** forms of the same plant

182 **filthy-mantled** covered with filthy scum. *Mantle* is both 'the green vegetable coating on standing water' and 'the foam that covers the surface of liquor' (*OED* 2c) —the latter sense is especially appropriate to the situation of the drunken conspirators.

There dancing up to th' chins, that the foul lake
O'er-stunk their feet.
PROSPERO   This was well done, my bird.
Thy shape invisible retain thou still.
The trumpery in my house, go bring it hither,
For stale to catch these thieves.
ARIEL   I go, I go.   *Exit*
PROSPERO
A devil, a born devil, on whose nature
Nurture can never stick; on whom my pains,
Humanely taken, all, all lost, quite lost;   190
And as with age his body uglier grows,
So his mind cankers. I will plague them all,
Even to roaring.
    Enter Ariel, loaden with glistering apparel, etc.
      Come, hang them on this line.
    Prospero and Ariel remain, invisible.
    Enter Caliban, Stephano, and Trinculo, all wet
CALIBAN Pray you tread softly, that the blind mole may not
hear a footfall. We now are near his cell.

193 *Enter . . . etc.*] CAPELL *(Re-enter . . .)*; *following* 'line' *in* F   193 them on] ROWE; *on them* F
193.1 *Prospero . . . invisible.*] CAPELL; *Prospero remains invisible* THEOBALD

183 **that** so that
184 **O'er-stunk their feet** Why only their
feet, if they are 'dancing up to th' chins'?
Bulloch suggested emending to 'feat', explaining that 'the filth they had passed
through had overstunk their nefarious
project', and Dover Wilson conjectured
'sweat'. Kermode comments, 'To be
precise, their feet when normally exposed,
not submerged' and concludes that 'there
is no point in trying to alter this'.
184 **bird** youngster; Prospero at 5.1.316
calls Ariel 'chick'.
186 **trumpery** attractive trash, the 'glistering apparel' of l. 193.1. Prospero, suiting
his art to its audience, produces carnival
costumes from his cell.
187 **stale** decoy
188–9 **nature | Nurture** The question of the
relation between these was a Renaissance
topos, and the ability of nurture—train-
ing, education—to transform nature in
any essential way was an ongoing subject
of debate. *As You Like It* includes a good
deal of inconclusive discussion of the
topic. In general, educational theorists

from Elyot to Bacon and Milton tended to
be optimistic, while literary and dramatic
texts, especially in the later sixteenth century, were ambivalent or pessimistic.
Proverbial wisdom taught that 'Nature
passes nurture' (Tilley N47); Prospero
analogously assumes that his education
has succeeded with Miranda and failed
with Caliban not through any defect in
the teaching methods but because his
pupils are respectively good and bad by
nature.
191–2 **as . . . cankers** The charge may be
less straightforward than it appears:
Prospero has just become conscious of his
own advancing age, and has expressed
fears for his own mind.
193 **line** a variant form of lind, the lime-tree
or linden. Although *line* could mean a
string, there are no contemporary
references to clothes-lines, and it is more
likely that the reference is to a stage-
property tree.
194–5 printed as verse by Rowe and most
subsequent editors

STEPHANO Monster, your fairy, which you say is a harmless
fairy, has done little better than played the jack with us.

TRINCULO Monster, I do smell all horse-piss, at which my
nose is in great indignation.

STEPHANO So is mine. Do you hear, monster? If I should   200
take a displeasure against you, look you—

TRINCULO Thou wert but a lost monster.

CALIBAN

Good my lord, give me thy favour still.
Be patient, for the prize I'll bring thee to
Shall hoodwink this mischance; therefore speak softly,
All's hushed as midnight yet.

TRINCULO Ay, but to lose our bottles in the pool!

STEPHANO There is not only disgrace and dishonour in that,
monster, but an infinite loss.

TRINCULO That's more to me than my wetting; yet this is   210
your harmless fairy, monster!

STEPHANO I will fetch off my bottle, though I be o'er ears for
my labour.

CALIBAN

Prithee, my king, be quiet. Seest thou here,
This is the mouth o' th' cell. No noise, and enter.
Do that good mischief which may make this island
Thine own forever, and I, thy Caliban,
For aye thy foot-licker.

STEPHANO Give me thy hand. I do begin to have bloody
thoughts.   220

TRINCULO O King Stephano! O peer! O worthy Stephano
—look what a wardrobe here is for thee!

197 **jack** knave (as in the deck of cards).
Kermode and Barton gloss 'jack o'lantern', or will-o'-the-wisp, but *OED* does
not record the word in this sense until
1673, and other Shakespearian uses of
'jack' give no support for such an interpretation. 'To play the jack with', meaning to deceive or make a fool of, was
proverbial (Tilley J8), and Benedick uses
the expression to Claudio: 'Do you play
the flouting jack, to tell us Cupid is a good
hare-finder . . . ?' (*Much Ado* 1.1.186).
204 **prize** booty
205 **hoodwink this mischance** render this
mischance harmless. To hoodwink is to

cover the eyes with a hood, or to blindfold.
206 **hushed as midnight** proverbial (Dent
M919.1)
212 **fetch off** either rescue or drink up
o'er ears i.e. drowned
221-2 The sight of the 'wardrobe' on the
lime-tree puts Trinculo in mind of a
ballad about King Stephen and clothing,
beginning

King Stephen was and a worthy peer,
His breeches cost him but a crown;
He held them sixpence all too dear,
With that he called the tailor lown
[= lout] . . . .

CALIBAN Let it alone, thou fool, it is but trash.

TRINCULO O ho, monster! We know what belongs to a frippery.

He takes a robe from the tree and puts it on
O King Stephano!

STEPHANO Put off that gown, Trinculo. (*Reaches for it*) By this hand, I'll have that gown.

TRINCULO Thy grace shall have it.

CALIBAN

The dropsy drown this fool! What do you mean   230
To dote thus on such luggage? Let't alone,
And do the murder first. If he awake,
From toe to crown he'll fill our skins with pinches,
Make us strange stuff.

STEPHANO Be you quiet, monster. Mistress line, is not this my jerkin? (*Removes it from the tree*) Now is the jerkin under the line. Now, jerkin, you are like to lose your hair, and prove a bald jerkin.

225.1 *He . . . on*] This edition; *not in* F   227 *Reaches for it*] This edition; *not in* F   231 Let't] RANN; Let's F   236 *Removes . . . tree*] This edition; *not in* F

The poem was later printed in Percy's *Reliques*, vol. 1 (1765), under the title 'Take thy old cloak about thee'. It is also quoted in *Othello* 2.3.92 ff.

224–5 **frippery** second-hand clothing shop (*OED* 3): i.e. this is *not* trash.

227 **Put off** take off

230 **dropsy** The disease is characterized by an excessive accumulation of fluid in the bodily tissues, and hence was used figuratively for an insatiable thirst or craving (*OED* 2).

231 **luggage** encumbrances (literally, what must be lugged about). Compare 5.1.298.

234 **stuff** referring both to the 'luggage' of l. 231 and the fabric of the 'glistering apparel'. The line should be read emphasizing '*us*'.

236–7 **Now . . . line** The literal sense is that Stephano has taken the jerkin off the tree. Attempts to explain the wordplay on *line* for the most part assume that 'under the line' means 'at the equator', as it does in *Henry VIII* 5.3.41–2, referring to a man with a fiery red nose: 'all that stand about him are under the line'. The allusion in l. 237–8 to losing one's hair then has to

do with fevers commonly contracted on long equatorial voyages, resulting in loss of hair, or possibly with having one's head shaved in the shipboard horseplay traditionally consequent on crossing the equator. Steevens suggested that a sexual pun is also involved, citing Rowley's *The Noble Soldier* (1634), in a carnal context, ''Tis hot going under the line there', and the anonymous *Lady Alimony* (c.1636–42; pub. 1659), 'Look to the clime | Where you inhabit; that's the torrid zone, | Yea, there goes the hair away.' Here the hair is lost through the usual treatment for syphilis (compare the discussion of the subject in *Errors* 2.2.68 ff.). Additional support for the conventional association of the equator with steamy sex is provided by R. A. Foakes in the Arden *Henry VIII*, citing Rowlands' *Knave of Hearts* (1613), '. . . consorted | With such hot spirited fiery feminine; | That heat him more than underneath the line' (p. 169). Finally, there may be an echo of the proverbial 'to strike the ball under the line' (Tilley B62), meaning not to play by the rules.

TRINCULO Do, do; we steal by line and level, an't like your
grace.   240
STEPHANO I thank thee for that jest: here's a garment for't.
 He takes a garment from the tree and gives it to Trinculo
Wit shall not go unrewarded while I am king of this
country. 'Steal by line and level' is an excellent pass of
pate.
 He takes another garment and gives it to him
There's another garment for't.
TRINCULO Monster, come put some lime on your fingers,
and away with the rest.
CALIBAN
I will have none on't. We shall lose our time,
And all be turned to barnacles, or to apes
With foreheads villainous low.   250
STEPHANO Monster, lay to your fingers. Help to bear this
away where my hogshead of wine is, or I'll turn you out
of my kingdom. Go to, carry this.
TRINCULO And this.
STEPHANO Ay, and this.
 *They give Caliban the remaining garments. A noise of
 hunters heard. Enter divers spirits in shape of dogs and
 hounds, hunting them about, Prospero and Ariel setting
 them on*
PROSPERO Hey, Mountain, hey!
ARIEL Silver! There it goes, Silver!

241.1 *He . . . Trinculo*] This edition; *not in* F   244.1 *He . . . him*] This edition; *not in* F
255.1 *They . . . garments*] This edition; *not in* F

239 **Do, do** expressing approval either of
Stephano's theft of the jerkin or of his
punning
 by line and level properly, according to
rule; the *line* and *level* are a plumb-line
and a carpenter's level. The expression
was proverbial (Tilley L305).
 an't like if it please
243–4 **pass of pate** skilful thrust; *pass* is a
fencing metaphor, *of pate* refers to 'the
head as the seat of the intellect; hence put
for skill, cleverness' (*OED* s. pate¹ 2).
246 **lime** birdlime, a sticky substance painted
on trees to catch birds. Thieves are said to
have sticky fingers; Tilley records 'His
fingers are lime twigs (F236).
249 **barnacles** Either the crustacean, or the

barnacle-goose popularly believed to
develop from it—the point may lie
precisely in its metamorphic quality. See
Gerard's *Herbal* (1597): 'There are . . . in
the north parts of Scotland . . . certain
trees, whereon do grow certain shell-
fishes . . . , which falling into the water, do
become fowls, whom we call barnacles, in
the north of England Brant [i.e. Brent]
geese, and in Lancashire tree geese' (cited
in the Variorum). A barnacle was also an
instrument of torture (*OED sb¹*), and thus
relevant to Caliban's fear of punishment
by pinches and cramps.
250 **villainous** vilely
251 **lay to** apply
 this the clothing from the lime-tree

PROSPERO Fury, Fury! There Tyrant, there! Hark, hark!
   Caliban, Stephano, and Trinculo are driven out
  Go charge my goblins that they grind their joints
  With dry convulsions, shorten up their sinews   260
  With agèd cramps, and more pinch-spotted make them
  Than pard or cat o' mountain.
ARIEL   Hark, they roar.
PROSPERO
  Let them be hunted soundly. At this hour
  Lies at my mercy all mine enemies.
  Shortly shall all my labours end, and thou
  Shalt have the air at freedom. For a little,
  Follow, and do me service.   *Exeunt*

5.1   *Enter Prospero in his magic robes, and Ariel*
PROSPERO
  Now does my project gather to a head.
  My charms crack not, my spirits obey, and Time
  Goes upright with his carriage. How's the day?

258.1 *Caliban ... out*] CAMBRIDGE; Calib. Steph. *and* Trinc. *driven out, roaring* THEOBALD; *not in* F

260 **dry convulsions** not recorded as a medical term; generally explained as relating to some presumed physiological theory about pains resulting from insufficient liquid in the body, or a deficiency in the humours.

261 **agèd cramps** the cramps of old age
  pinch-spotted black and blue from pinches

262 **pard, cat o' mountain** Both terms applied to the leopard or panther, and the latter in addition to various kinds of wildcat. Of these animals, however, only the leopard is spotted. Noble (p. 251) suggests that this may be a second allusion to the Bishops' Bible version of Jeremiah 13: 23: 'May a man of Ind change his skin, and the cat of the mountain her spots?' Compare 2.2.57.

264 **Lies** For the singular verb preceding a plural subject, see Abbott 335.

267 **Exeunt** The fact that Prospero and Ariel leave the stage and are onstage again at the opening of Act 5 was thought by Dover Wilson to be evidence of a cut. Greg objected that it shows, on the contrary, 'that the act division is original' (*The*

Shakespeare First Folio, p. 418, n. 1). It also implies that an interval separated the acts.

5.1.1 **project** design, scheme; and, taken with *gather to a head* and *crack* (l. 2), suggesting an alchemical metaphor. Sir Epicure Mammon uses the word when he believes he is about to get possession of the philosophers' stone: 'My only care is | Where to get stuff enough now to project on' (Ben Jonson, *The Alchemist*, 2.2.11–12). Projection is 'the casting of the powder of philosophers' stone (*powder of projection*) upon a metal in fusion to effect its transmutation into gold' (*OED* 2).
  gather to a head approach a crisis; perhaps more specifically come to a boil, as liquid, continuing the projection metaphor

2 **crack** fail; Kermode points out that the word 'is used of the explosion of retorts, etc., which brings Mammon's venture to disaster in Jonson's *Alchemist* (4.5.56)'.

3 **Goes ... carriage** walks without stooping because his burden (*carriage* = what he carries) is no longer heavy

ARIEL
   On the sixth hour, at which time, my lord,
   You said our work should cease.
PROSPERO   I did say so
   When first I raised the tempest. Say, my spirit,
   How fares the King and's followers?
ARIEL   Confined together
   In the same fashion as you gave in charge,
   Just as you left them; all prisoners, sir,
   In the line-grove which weather-fends your cell;   10
   They cannot budge till your release. The King,
   His brother, and yours, abide all three distracted,
   And the remainder mourning over them,
   Brimful of sorrow and dismay; but chiefly
   Him that you termed, sir, the good old Lord Gonzalo,
   His tears runs down his beard like winter's drops
   From eaves of reeds. Your charm so strongly works 'em
   That if you now beheld them, your affections
   Would become tender.
PROSPERO   Dost thou think so, spirit?
ARIEL
   Mine would, sir, were I human.
PROSPERO   And mine shall.   20
   Hast thou, which art but air, a touch, a feeling
   Of their afflictions, and shall not myself,
   One of their kind, that relish all as sharply
   Passion as they, be kindlier moved than thou art?
   Though with their high wrongs I am struck to th' quick,

5.1.20 human] F (humane)   23 sharply] F3; sharply, F1

10 **line-grove** lime or linden grove; see
   4.1.193.
   weather-fends protects you against the
   weather
11 **your release** you release them
16 **tears runs** For the syntax, see Abbott
   333.
17 **eaves of reeds** thatched roofs
18 **affections** feelings
21 **touch** delicate perception
23–4 **that … they** If F's comma after *sharply*
   is correct, *all* is an adverb and *passion* is a
   verb—there are analogues for the latter
   in *LLL* 1.1.260 and *Two Gentlemen*
   4.4.167, and this is the reading supported

by the *OED*. In this case, *Passion as they* is
an extension of *relish all as sharply*: Prospero says, 'I enjoy everything as keenly as
they, I am as affected by deep feeling.'
Editors, however, have almost without
exception preferred F3's emendation,
which gives the stronger sense: 'I who
am fully as sensitive to suffering as they
are' (but as Ariel is not).
24 **kindlier** both more generously and more
humanly, in accordance with my kind
25 **quick** the tenderest or most vital part. 'To
touch to the quick' remains proverbial
(see Tilley Q13).

Yet with my nobler reason 'gainst my fury
Do I take part. The rarer action is
In virtue than in vengeance. They being penitent,
The sole drift of my purpose doth extend
Not a frown further. Go, release them, Ariel.   30
My charms I'll break, their senses I'll restore,
And they shall be themselves.

ARIEL   I'll fetch them, sir.   *Exit*
⌜ *Prospero traces a magic circle on the stage with his
    staff* ⌝

PROSPERO
Ye elves of hills, brooks, standing lakes, and groves,
And ye that on the sands with printless foot
Do chase the ebbing Neptune, and do fly him
When he comes back; you demi-puppets that
By moonshine do the green sour ringlets make,
Whereof the ewe not bites; and you whose pastime
Is to make midnight mushrooms, that rejoice
To hear the solemn curfew, by whose aid—   40

32.1–2 *Prospero . . . staff*] WILSON; *not in* F

27-8 **The . . . vengeance** Like 2.1.145 ff.,
this is closely related to a passage in
Florio's Montaigne (first noted by Eleanor
Prosser, 'Shakespeare, Montaigne and
the "Rarer Action"', *Shakespeare Studies*
I (1961), pp. 261–6; see Appendix D).
Compare also Sonnet 94: 'They that have
power to hurt and will do none | . . . right-
ly do inherit heaven's graces'. Behind
both Shakespeare and Montaigne is
proverbial wisdom: 'To be able to do
harm and not do it is noble' (Tilley H170).
28 **virtue** a much broader concept than the
expected 'forgiveness'. *Virtue* is parallel
with and a logical extension of *reason* in
l. 26. The classical *virtus* implied both
heroic magnanimity and the stoic ability
to remain unmoved by suffering.
  They . . . penitent These conditions are
not met. See the Introduction, pp. 51 ff.
32.1 *Prospero traces a magic circle* F at
l. 57.5 has the shipwreck victims 'all
enter the circle which Prospero had
made', but gives no indication of when he
makes it. Either here or when he begins
his 'airy charm' at l. 52 would be
appropriate points.
33-50 **Ye elves . . . art** This passage is a close
translation of a speech of Medea's in the

Metamorphoses vii. 197–209. Shakespeare seems to have used both the original and Golding's Elizabethan translation (see Appendix E). The syntax is
'Ye elves . . . by whose aid I have bedimmed . . . , called forth . . . , and . . . set;
. . . given fire, and rifted . . . '. The passage
ends without a main verb.
35 **fly** flee
36 **demi-puppets** because they are partially
subject to Prospero's will, or because they
are diminutive dolls; in either case, with a
theatrical overtone
37 **green sour ringlets** fairy rings, small
circles of sour grass near, and caused by
the roots of, toadstools. They were said to
be caused by fairies dancing.
39 **midnight mushrooms** mushrooms that
appear overnight
  that you who
40 **curfew** the evening bell, rung at nine
o'clock in Shakespeare's time, originally
as a signal to extinguish all household
fires, but by the sixteenth century merely
marking the hour. Spirits were said to be
free to walk either then or at midnight: in
Lear 3.4.115–16, 'the foul fiend Flibber-
tigibbet', according to Edgar, 'begins at
curfew and walks till the first cock'.

Weak masters though ye be—I have bedimmed
The noontide sun, called forth the mutinous winds,
And 'twixt the green sea and the azured vault
Set roaring war; to the dread rattling thunder
Have I given fire, and rifted Jove's stout oak
With his own bolt; the strong-based promontory
Have I made shake, and by the spurs plucked up
The pine and cedar. Graves at my command
Have waked their sleepers, oped, and let 'em forth
By my so potent art. But this rough magic   50
I here abjure; and when I have required
Some heavenly music—which even now I do—
To work mine end upon their senses that
This airy charm is for, I'll break my staff,
Bury it certain fathoms in the earth,
And deeper than did ever plummet sound
I'll drown my book.

41 **masters** ministers or instruments: compare *Macbeth* 4.1.63, where the witches refer to the apparitions they have summoned up as 'our masters'. Kermode cites *Faerie Queene* III viii. 4: 'She was wont her sprites to entertain, | The masters of her art', along with a Jonson passage that is not relevant. For a discussion of the etymology and development of the term, which is cognate with *mystery* in the sense of 'craft', see M. K. Flint and E. J. Dobson, 'Weak Masters', *RES* NS 10 (1959), pp. 58–60.

45 **rifted** split
   Jove's . . . oak The tree was sacred to Jove because of its hardness and endurance.

47 **spurs** roots (*OED* 9)

48–9 **Graves . . . forth** This claim is made nowhere else in the play; it expands Medea's 'manesque exire sepulcris' (in Golding, 'I call up dead men from their graves'). Cerimon, in *Pericles*, who restores Thaisa to life, gives cautious assent to the ability of natural magic to raise the dead: 'Death may usurp on nature many hours, | And yet the fire of life kindle again' (3.2.85–6); but this is a far more modest claim than Prospero's or Medea's.

For graves opening and releasing their inhabitants, compare *Dream* 5.1.379–81: 'Now it is the time of night | When the graves all gaping wide, | Every one lets forth his sprite . . . ', and Horatio's report that in Rome 'A little ere the mightiest Julius fell, | The graves stood tenantless and the sheeted dead | Did squeak and gibber in the Roman streets' (*Hamlet* 1.1.114–16).

50 **rough magic** The renunciation of the *potent art* is manifest in Prospero's language. Barton argues that the phrase suggests a new awareness of the limitations of Prospero's powers: he can raise storms, but cannot enforce penitence or changes of heart (Introduction, p. 29). Kermode's explanation, 'unsubtle by comparison with the next degree of the mage's enlightenment', requires the dubious assumption that Prospero is pursuing a systematic Neoplatonic ascent. See the Introduction, p. 23.

53 **their senses that** the senses of those who

54 **airy** in the air, with an overtone of the musical sense of *air*, 'song-like music, melody' (*OED* iv. 18)

Solemn music.
Here enters Ariel before; then Alonso, with a frantic
gesture, attended by Gonzalo; Sebastian and Antonio
in like manner, attended by Adrian and Francisco.
They all enter the circle which Prospero had made, and
there stand charmed; which Prospero observing,
speaks

A solemn air, and the best comforter
To an unsettled fancy, cure thy brains,
Now useless, boil within thy skull. There stand,　　60
For you are spell-stopped.
Holy Gonzalo, honourable man,
Mine eyes, ev'n sociable to the show of thine,
Fall fellowly drops. The charm dissolves apace,
And as the morning steals upon the night,
Melting the darkness, so their rising senses
Begin to chase the ignorant fumes that mantle
Their clearer reason. O good Gonzalo,
My true preserver, and a loyal sir
To him thou follow'st, I will pay thy graces　　70
Home both in word and deed! Most cruelly
Didst thou, Alonso, use me and my daughter.
Thy brother was a furtherer in the act—

---

60 boil] F (boile); boil'd ROWE　72 Didst] F *catchword*; did F

57.6–7 **Prospero . . . speaks** Prospero is invisible and inaudible to the assembled
company until he reveals himself at l.
106.
58–9 **A solemn . . . fancy** Musical therapy
had been in use since ancient times as a
way of adjusting the distempered harmony of the human system. Compare
Lorenzo's explanation of the Orpheus
legend to Jessica in *Merchant* 5.1.70 ff.
59 **unsettled fancy** disturbed imagination,
here with implications of delusion and in-
sanity
59–60 **thy . . . thy** Prospero begins his
address to Alonso as leader of the party.
60 **Now useless, boil** Most editors, following
Rowe, emend F's 'boile' to 'boiled', but
the change is both unnecessary and mis-
leading: the emendation puts the action
in the past, but the torture is still going on.

The construction is another *apo koinu*, or
grammatical zeugma; the sense is, 'thy
brains . . . which boil'. Compare 3.3.77.
See Abbott 244 (omission of the relative).
63 **sociable** humanly sympathetic
show appearance (Gonzalo is weeping)
64 **Fall** let fall
67 **ignorant . . . mantle** fogs that keep
in ignorance. Compare *Winter's Tale*
1.2.394–6: 'If you know aught which
does behoove my knowledge | Thereof to
be informed, imprison't not | In ignorant
concealment.'
68 **clearer** growing clearer: '*Clearer* may be
regarded as proleptic, having the force of
"and thus make it clearer"' (Kermode).
69 **sir** gentleman
70–1 **pay thy graces | Home** fully repay thy
favours. 'To pay home' was proverbial
(Dent H535.1).

Thou art pinched for't now, Sebastian! Flesh and
    blood,
You, brother mine, that entertained ambition,
Expelled remorse and nature, whom, with Sebastian—
Whose inward pinches therefore are most strong—
Would here have killed your king, I do forgive thee,
Unnatural though thou art.—Their understanding
Begins to swell, and the approaching tide   80
Will shortly fill the reasonable shore,
That now lies foul and muddy. Not one of them
That yet looks on me, or would know me. Ariel,
Fetch me the hat and rapier in my cell.
        Exit Ariel and returns immediately
I will discase me, and myself present
As I was sometime Milan. Quickly, spirit!
Thou shalt ere long be free.
        Ariel sings, and helps to attire him

---

75 entertained] F2 (entertain'd); entertaine F1   82 lies] F3; ly F1   84.1 *Exit . . . immediately*]
CAPELL; *after l.86* in THEOBALD

74 **pinched** tortured, punished. The
    metaphor is literalized in Prospero's
    threats to Caliban: 'Thou shalt be
    pinched | As thick as honeycomb,'
    1.2.328–9.
75 **entertained** Editors from F2 on have thus
    emended F's 'entertaine'. The parallelism
    of *expelled* immediately following (l. 76)
    suggests that this is correct, though since
    Antonio is unrepentant, a case could be
    made for retaining the present tense.
    (Compare 'boil', l. 60.)
76 **remorse** pity
    whom emended by Rowe and many
    others to 'who'. Kermode claims, citing
    Onions, that the construction is not ex-
    traordinary, but the *Shakespeare Glossary*
    in fact gives no support for this: all
    Onions's examples except this one involve
    a confusion of two constructions (e.g.
    'Arthur, whom they say is killed tonight',
    King John 4.2.165). Prospero's usage can-
    not be explained in this way. If the
    passage is evidence of anything other
    than an error, it is not of normal Shake-
    spearian syntax but of Prospero's anxiety
and uncertainty—or Shakespeare's on
his behalf—as he faces the difficulty of
actually forgiving his brother. Alonso
produces a comparable example at l. 136,
almost always emended, from F2 on, to
'who'; both show Shakespeare abruptly
changing direction in mid-sentence.
79–82 **understanding . . . muddy** The
metaphor elaborates the notion of their
minds returning to reason: understanding is conceived as the sea and the uncom-
prehending reason as the shore at low
tide, which will be cleared of its obscuring
scum and mud (compare 'the ignorant
fumes' of l. 67) as the tide turns.
81 **reasonable shore** shore of reason
82 **Not one of them** 'There is' understood
84 **hat and rapier** These are essential elements of aristocratic dress.
85 **discase me** take off my magic robes.
Discase (for 'undress') is used only by
Shakespeare, though 'uncase' is common
in this sense.
86 **sometime Milan** formerly, when Duke of
Milan

ARIEL

    Where the bee sucks, there suck I,
    In a cowslip's bell I lie;
    There I couch when owls do cry;   90
    On the bat's back I do fly
    After summer merrily.
    Merrily, merrily shall I live now
    Under the blossom that hangs on the bough.

PROSPERO

Why, that's my dainty Ariel! I shall miss thee,
But yet thou shalt have freedom. (*Arranging his attire*)
               So, so, so.
To the King's ship, invisible as thou art;
There shalt thou find the mariners asleep
Under the hatches. The master and the boatswain
Being awake, enforce them to this place,   100
And presently, I prithee.

ARIEL

I drink the air before me, and return
Or ere your pulse twice beat.   *Exit*

GONZALO

All torment, trouble, wonder, and amazement
Inhabits here. Some heavenly power guide us
Out of this fearful country!

PROSPERO   Behold, sir King,
The wrongèd Duke of Milan, Prospero.
For more assurance that a living prince
Does now speak to thee, I embrace thy body,
    Embraces Alonso

---

96 *Arranging his attire*] This edition; *not in* F   109.1 *Embraces Alonso*] This edition; *not in* F

88–94 The song is Ariel's proleptic celebration of freedom. For the music by Robert Johnson, see Appendix C.

92 **After summer** pursuing summer, as birds migrate when the weather grows cold. Ariel anticipates a life of everlasting summer, as Prospero's masque had promised the lovers a world without winter.

98–9 **There . . . hatches** Prospero reports to Ariel information he has had from Ariel in the first place: see 1.2.230.

101 **presently** immediately

102 **I drink the air** The Latinism is *viam vorare*, to devour the road. In 2 *Henry IV* a gentleman fleeing from Percy's rebellion 'seemed in running to devour the way' (1.1.47); Ariel adapts the expression to his own mode of travel. Hamlet's 'I eat the air, promise-crammed' (3.2.94), i.e. I live on promises, may also be relevant.

103 **Or ere** Both words mean 'before'; compare 1.2.11.

105 **Inhabits** For the construction, see Abbott 333.

And to thee and thy company I bid   110
A hearty welcome.
ALONSO   Whe'er thou beest he or no,
Or some enchanted trifle to abuse me,
As late I have been, I not know. Thy pulse
Beats as of flesh and blood; and since I saw thee,
Th'affliction of my mind amends, with which
I fear a madness held me. This must crave,
An if this be at all, a most strange story.
Thy dukedom I resign, and do entreat
Thou pardon me my wrongs. But how should Prospero
Be living, and be here?
PROSPERO (*to Gonzalo*) First, noble friend,   120
Let me embrace thine age, whose honour cannot
Be measured or confined.
          Embraces Gonzalo
GONZALO   Whether this be,
Or be not, I'll not swear.
PROSPERO   You do yet taste
Some subtleties o'th' isle, that will not let you
Believe things certain. Welcome, my friends all!
(*Aside to Sebastian and Antonio*) But you, my brace of
          lords, were I so minded,
I here could pluck his highness' frown upon you,
And justify you traitors. At this time
I will tell no tales.
SEBASTIAN (*aside*)   The devil speaks in him!

---

111 Whe'er] CAPELL; Where F   122 *Embraces Gonzalo*] This edition; *not in* F   124 not] F3;
nor F1

111 **Whe'er** for 'whether'
112 **enchanted   trifle**   insubstantial   ap-
     parition produced by magic
     abuse both delude and maltreat
113 **have been** i.e. abused
116 **crave** call for
117 **An . . . all** if this is not another illusion
119 **my wrongs** the wrongs I have done you
124 **subtleties** deceptions, illusions, and
     with 'taste', punning on the sense of

elaborate ornamental sugar confections
arranged as a pageant and served at the
conclusion of a banquet
126 **brace** pair
128 **justify** prove
129 **no** most likely a repetition of his deter-
     mination to 'tell no tales'. If it is an indica-
     tion that Prospero has heard Sebastian's
     remark, his answer is a very weak one.

PROSPERO   No.
   For you, most wicked sir, whom to call brother   130
   Would even infect my mouth, I do forgive
   Thy rankest fault—all of them—and require
   My dukedom of thee, which perforce I know
   Thou must restore.
ALONSO   If thou beest Prospero,
   Give us particulars of thy preservation,
   How thou hast met us here, whom three hours since
   Were wrecked upon this shore, where I have lost—
   How sharp the point of this remembrance is!—
   My dear son Ferdinand.
PROSPERO   I am woe for't, sir.
ALONSO
   Irreparable is the loss, and patience   140
   Says it is past her cure.
PROSPERO   I rather think
   You have not sought her help, of whose soft grace
   For the like loss I have her sovereign aid,
   And rest myself content.
ALONSO   You the like loss?
PROSPERO
   As great to me, as late; and supportable
   To make the dear loss, have I means much weaker
   Than you may call to comfort you, for I
   Have lost my daughter.
ALONSO   A daughter?
   O heavens, that they were living both in Naples,
   The king and queen there! That they were, I wish   150
   Myself were mudded in that oozy bed
   Where my son lies. When did you lose your daughter?

---

136 whom] F1; who F2, ROWE, etc. (see note to 5.1.76)

136 **whom** See l. 76.
139 **woe** sorry
142 **of** by
145 **late** recent
146 **dear** grievous: compare 2.1.133
150 **That** provided that (i.e. if my death

would bring it about). For the construction, see Abbott 364.
151 **mudded** Compare 3.3.102: these are
OED's only instances of the sense 'buried
in mud'.

PROSPERO

    In this last tempest. I perceive these lords
At this encounter do so much admire
That they devour their reason, and scarce think
Their eyes do offices of truth, their words
Are natural breath; but howsoe'er you have
Been jostled from your senses, know for certain
That I am Prospero, and that very Duke
Which was thrust forth of Milan, who most strangely   160
Upon this shore, where you were wrecked, was landed
To be the lord on't. No more yet of this,
For 'tis a chronicle of day by day,
Not a relation for a breakfast, nor
Befitting this first meeting. Welcome, sir;
This cell's my court. Here have I few attendants,
And subjects none abroad. Pray you look in.
My dukedom since you have given me again,
I will requite you with as good a thing,
At least bring forth a wonder to content ye   170
As much as me my dukedom.

    Here Prospero discovers Ferdinand and Miranda
    playing at chess

156 truth. their] ROWE; Truth: Their F

154 **encounter** meeting
    admire wonder
154–5 **do . . . reason** are so amazed that they are at a loss for rational discourse; their reason is swallowed up in amazement.
156 **do offices of truth** function accurately
156–7 **their . . . breath** If F's punctuation is interpreted according to modern usage, this is an independent clause preceded by a colon, in which case it means that their words are merely breath, not rational discourse. Capell, followed in this century only by Dover Wilson, retained the colon and emended *their* to 'these', assuming that the lords' difficulty was in believing the reality of Prospero's words, not their own. But F often separates clauses with a colon where modern usage would employ a comma, and the most likely reading is the one adopted by Rowe and followed by most editors, in which the passage is a second clause dependent on *think*, and the point is that they cannot credit their words either because they believe they are

dreaming, or because their expressions of astonishment are forced from them.
160 **strangely** wonderfully
163 **of day by day** to be told over many days
164 **relation** story
167 **abroad** elsewhere, beyond what is right here
170 **bring forth a wonder** Prospero, punning on Miranda's name, seems to promise another illusion.
171.1 *discovers* reveals (by pulling aside a curtain). If the discovery place is Prospero's cell—into which the Neapolitans have just been invited to look—Ferdinand and Miranda must be concealed elsewhere, possibly in a small portable pavilion placed on stage only for this scene. Alternatively, if they are in the discovery place, one of the stage doors may be doing service for Prospero's cell, which would be established as such by his entrance through it in his magic robes at the opening of the scene.

MIRANDA

Sweet lord, you play me false.

FERDINAND   No, my dearest love,

I would not for the world.

MIRANDA

Yes, for a score of kingdoms you should wrangle,

And I would call it fair play.

ALONSO   If this prove

A vision of the island, one dear son

Shall I twice lose.

SEBASTIAN   A most high miracle!

FERDINAND (*coming forward*)

Though the seas threaten, they are merciful.

I have cursed them without cause.

    He kneels before Alonso

ALONSO   Now all the blessings

Of a glad father compass thee about!   180

Arise, and say how thou cam'st here.

    Ferdinand rises

MIRANDA   O wonder!

How many goodly creatures are there here!

---

178 *coming forward*] This edition; *not in* F   179 *He . . . Alonso*] THEOBALD (*Kneels*); *not in* F

171.2 **playing at chess** The territorial ambitions of their elders are transformed by Ferdinand and Miranda into the stratagems of chess. (See the Introduction, pp. 29–30.) The game was an aristocratic pastime associated especially with lovers, often with illicit sexual overtones, and also served as a frequent allegory of politics, most notoriously in Middleton's *Game at Chess* (1621). The traditional implications of the game in relation to this scene are discussed by Bryan Loughrey and Neil Taylor, 'Ferdinand and Miranda at Chess', *Shakespeare Survey* 35 (Cambridge, 1982), pp. 113–18.

172 **You play me false** Though most editors have resisted the idea, Miranda is clearly accusing Ferdinand of cheating here, and in ll. 174–5 declares her willing complicity in the act (see the Introduction, pp. 29–30). The remark is an affectionate echo of Prospero's earlier charge of

treason against Ferdinand (1.2.454 ff.).

174–5 'His hyperbolic assertion that he would not play her false she then proceeds to deflate . . . with the remark that for such a stake as a score of kingdoms he certainly would do so, and that she, in her love for him, would call it fair' (Barton).

174 **wrangle** contend with me, be my adversary. The word has no necessary imputation of dishonest dealing, but it is here presented as an antithesis to *fair play* (l. 175), and dishonesty must be implied. The stake in the larger context of their 'wrangling' is, of course, Milan, the realm he has played for and won.

175 **And** Possibly this is the variant form of 'an' (= if), in which case Miranda is making her approval a necessary condition of Ferdinand's playing false.

176 **A vision** another illusion

177 **A . . . miracle** Sebastian is either for once impressed, or merely being characteristically sarcastic.

How beauteous mankind is! O brave new world
That has such people in't!
PROSPERO   'Tis new to thee.
ALONSO
What is this maid with whom thou wast at play?
Your eld'st acquaintance cannot be three hours.
Is she the goddess that hath severed us,
And brought us thus together?
FERDINAND   Sir, she is mortal;
But by immortal providence, she's mine.
I chose her when I could not ask my father   190
For his advice—nor thought I had one. She
Is daughter to this famous Duke of Milan,
Of whom so often I have heard renown,
But never saw before; of whom I have
Received a second life; and second father
This lady makes him to me.
ALONSO   I am hers.
But O, how oddly will it sound that I
Must ask my child forgiveness!
PROSPERO   There, sir, stop.
Let us not burden our remembrances with
A heaviness that's gone.
GONZALO   I have inly wept,   200
Or should have spoke ere this: look down, you gods,
And on this couple drop a blessèd crown;
For it is you that have chalked forth the way
Which brought us hither.

---

183 **mankind** presumably humanity; but,
other than Miranda, there are only men
present.
186 **eld'st** longest
187 **Is she the goddess** Compare Ferdinand's
response to his first sight of Miranda,
1.2.422.
196 **I am hers** I am her servant; my respects
to her
199 **remembrances** From Rowe on, regularly emended to 'remembrance' until Allen
(*Notes*, p. 64) suggested, in accordance
with a principle of Walker's, that the -s
was elided. Kermode prints 'remembrance'. However, none of the four other

Shakespearian examples of the word is
elided (e.g. 'By our remembrances of days
foregone,' *All's Well* 1.3.134), and the
choice here is whether or not to emend
purely in the interests of metrical regu-
larity. Since Shakespeare elsewhere invariably uses the word in the plural with
a plural subject, I have assumed that F's
reading is correct.
200 **heaviness** grief
203 **chalked forth the way** marked the true
path as with a chalk line. Compare *Henry
VIII* 1.1.60, 'ancestry, whose grace |
Chalks successors their way . . .'.

ALONSO   I say 'amen', Gonzalo.
GONZALO
Was Milan thrust from Milan that his issue
Should become kings of Naples? O rejoice
Beyond a common joy, and set it down
With gold on lasting pillars! In one voyage
Did Claribel her husband find at Tunis,
And Ferdinand, her brother, found a wife   210
Where he himself was lost, Prospero his dukedom
In a poor isle, and all of us ourselves
When no man was his own.
ALONSO (*to Ferdinand and Miranda*)
                                Give me your hands.
Let grief and sorrow still embrace his heart
That doth not wish you joy!
GONZALO   Be it so, amen.
        Enter Ariel, with the Master and Boatswain amazedly
        following
O look, sir, look, sir, here is more of us!
I prophesied if a gallows were on land
This fellow could not drown. (*To Boatswain*) Now,
        blasphemy,
That swear'st grace o'erboard, not an oath on shore?
Hast thou no mouth by land? What is the news?   220
BOATSWAIN
The best news is that we have safely found

---

218 *To Boatswain*] This edition; *not in* F

---

205 **Milan . . . Milan** the Duke of Milan . . .
    the realm
208 **on lasting pillars** There may be an
    allusion here to the imperial emblem of
    Charles V, the pillars of Hercules, familiar
    from the triumphal iconography of the
    Holy Roman and the Spanish Empires,
    and subsequently adapted by monarchs
    throughout Europe, as well as by
    Elizabeth I after the defeat of the Armada.
    For a brief summary of the iconographic
    tradition, see Dennis C. Kay, 'Gonzalo's
    "Lasting Pillars": *The Tempest*, v.i.208',
    ShQ, 35 (1984), pp. 322–4.
213 **When . . . own** when we had lost

our senses. Gonzalo's characteristically
optimistic summary ignores the un-
regeneracy of Antonio and Sebastian.
214 **still** forever
218 **blasphemy** quintessential blasphemer,
the embodiment of blasphemy (compare
'Bravely, my diligence', l. 241). It was
Sebastian who had accused the Boatswain of blasphemy, 1.1.40.
219 **swear'st grace o'erboard** drivest God's
grace from the ship by swearing. Kermode, citing R. R. Cawley, notes that
'Ralegh ordered his captains to "take
especial care that God be not blasphemed
in your ship"'.

Our king and company; the next, our ship,
Which but three glasses since we gave out split,
Is tight and yare and bravely rigged as when
We first put out to sea.
ARIEL (*aside to Prospero*)   Sir, all this service
Have I done since I went.
PROSPERO (*aside to Ariel*)   My tricksy spirit!
ALONSO
These are not natural events, they strengthen
From strange to stranger. Say, how came you hither?
BOATSWAIN
If I did think, sir, I were well awake,
I'd strive to tell you. We were dead of sleep,   230
And—how we know not—all clapped under hatches,
Where but even now with strange and several noises
Of roaring, shrieking, howling, jingling chains,
And more diversity of sounds, all horrible,
We were awaked, straightway at liberty,
Where we, in all our trim, freshly beheld
Our royal, good, and gallant ship; our master
Cap'ring to eye her—on a trice, so please you,
Even in a dream, were we divided from them,
And were brought moping hither.
ARIEL (*aside to Prospero*)   Was't well done?   240
PROSPERO (*aside to Ariel*)
Bravely, my diligence. Thou shalt be free.
ALONSO
This is as strange a maze as e'er men trod,

241 Bravely . . . Thou] F4; Bravely (my diligence) thou F1

223 **three glasses since** three hours ago; see
l. 186 and 1.2.240.
gave out declared
224 **yare** trim, seaworthy
226 **tricksy** playful, sportive; also ingenious. Onions suggests 'full of devices,
resourceful'.
227 **strengthen** increase
230 **of** with
232 **several** diverse

236–7 **our trim . . . ship** i.e. our garments,
like our ship, were undamaged
238 **Cap'ring** dancing for joy
on a trice in an instant. The expression
was conventional (compare Tilley T517).
239 **them** the other members of the crew
240 **moping** bewildered
242 **maze** Compare Gonzalo's characterization of the island: 'Here's a maze trod
indeed,' 3.3.2.

And there is in this business more than nature
Was ever conduct of. Some oracle
Must rectify our knowledge.
PROSPERO   Sir, my liege,
Do not infest your mind with beating on
The strangeness of this business. At picked leisure,
Which shall be shortly single, I'll resolve you,
Which to you shall seem probable, of every
These happened accidents; till when, be cheerful   250
And think of each thing well. (*Aside to Ariel*) Come
      hither, spirit.
Set Caliban and his companions free;
Untie the spell.   *Exit Ariel*
            How fares my gracious sir?
There are yet missing of your company
Some few odd lads that you remember not.
      Enter Ariel, driving in Caliban, Stephano, and Trinculo
      in their stolen apparel
STEPHANO Every man shift for all the rest, and let no man
      take care for himself, for all is but fortune. Coraggio,
      bully-monster, coraggio!
TRINCULO If these be true spies which I wear in my head,
      here's a goodly sight.   260

---

253 *Exit Ariel*] CAPELL; *not in* F   258 Coraggio] F2 (Coragio); Corasio F1

---

244 **conduct** director: compare *Romeo*
5.3.116, 'Come, bitter conduct, come,
unsavoury guide'
246 **infest** trouble
beating hammering, insistently thinking
(compare 1.2.176)
247 **picked leisure** leisure time that we shall
select
248 **Which . . . single** which (i.e. leisure) will
soon be continuous: Prospero amends his
previous remark. The phrase in F is in parentheses. Rowe, followed by many editors,
emended the line to 'Which shall be shortly,
single [i.e. privately] I'll resolve you . . . .'
249 **Which . . . probable** and my explanation
will convince you; *probable* = capable of
being proved (*OED* 1)
every each of (Abbott 12)
250 **accidents** events that have occurred
(Abbott 295)

251 **well** favourably
253 **Untie the spell** Enchanted characters
are regularly described in the play as 'knit
up' (Alonso, Sebastian, etc., 3.3.89) or
'bound' (Ferdinand, 1.2.487).
256–7 **Every . . . himself** another of
Stephano's drunken reversals; compare
3.2.126–7.
257, 258 **Coraggio** courage (Italian); possibly an affected usage, or perhaps a little
Neapolitan local colour. *OED* compares
'bravo' as a hortatory exclamation.
258 **bully** 'A term of endearment and
familiarity . . . implying friendly admiration . . . . Often prefixed as a sort of title to
the name or designation of the person
addressed, as in "bully Bottom", "bully
doctor"' (*OED* 1).
259 **If . . . head** if my eyes can be trusted

CALIBAN

O Setebos, these be brave spirits indeed.
How fine my master is! I am afraid
He will chastise me.

SEBASTIAN   Ha, ha!

What things are these, my lord Antonio?
Will money buy 'em?

ANTONIO   Very like. One of them

Is a plain fish, and no doubt marketable.

PROSPERO

Mark but the badges of these men, my lords,
Then say if they be true. This misshapen knave,
His mother was a witch, and one so strong
That could control the moon, make flows and ebbs,   270
And deal in her command without her power.
These three have robbed me, and this demi-devil—
For he's a bastard one—had plotted with them
To take my life. Two of these fellows you
Must know and own; this thing of darkness I
Acknowledge mine.

CALIBAN   I shall be pinched to death!

ALONSO Is not this Stephano, my drunken butler?

SEBASTIAN He is drunk now—where had he wine?

ALONSO

And Trinculo is reeling-ripe! Where should they
Find this grand liquor that hath gilded 'em?   280
How cam'st thou in this pickle?

261 **Setebos** Caliban's god: see 1.2.372.
  brave handsome, impressive
262 **fine** splendidly dressed: Prospero is in
  his ducal garments (see ll. 85–6).
266 **fish** recalling Trinculo's doubt about
  whether Caliban was more like a man or
  a fish: see 2.2.24 ff.
267 **badges** heraldic devices worn as identification by the servants of great houses.
  The point is that, though they are Alonso's servants, their 'livery' has been
  stolen from Prospero.
268 **true** honest
270 **control the moon** as Medea does,
  Metamorphoses vii. 207
  make flows and ebbs as the moon does
271 **without** beyond the reach of: compare

Dream 4.1.159, 'Without the peril of
th'Athenian law.'
272 **demi-devil** recalling the charge that
  Caliban was 'got by the devil himself |
  Upon thy wicked dam' (1.2.319–20).
  Both may be mere invective, as Othello
  calls Iago a demi-devil, 5.2.300. See the
  Introduction, p. 25.
275 **own** acknowledge
279 **reeling-ripe** ready (i.e. drunk enough)
  for reeling
280 **gilded** flushed their faces (*OED* s.
  gild v¹. 6)
281 **pickle** both 'preserving liquor' and
  'predicament'. 'To be in a pickle' remains
  proverbial (Tilley P276).

TRINCULO I have been in such a pickle since I saw you last
that I fear me will never out of my bones. I shall not fear
fly-blowing.

SEBASTIAN Why, how now, Stephano?

STEPHANO
O, touch me not; I am not Stephano, but a cramp.

PROSPERO
You'd be king o' the isle, sirrah?

STEPHANO
I should have been a sore one then.

ALONSO (*indicating Caliban*)
This is a strange thing as e'er I looked on.

PROSPERO
He is as disproportioned in his manners   290
As in his shape. Go, sirrah, to my cell;
Take with you your companions. As you look
To have my pardon, trim it handsomely.

CALIBAN
Ay, that I will; and I'll be wise hereafter,
And seek for grace. What a thrice-double ass
Was I to take this drunkard for a god,
And worship this dull fool!

PROSPERO   Go to, away.

ALONSO
Hence, and bestow your luggage where you found it.

SEBASTIAN Or stole it, rather.

                    Exeunt Caliban, Stephano, and Trinculo

PROSPERO
Sir, I invite your highness and your train   300
To my poor cell, where you shall take your rest
For this one night, which part of it I'll waste

---

289 *indicating Caliban*] STEEVENS (*pointing to Caliban*); *not in* F   299.1 *Exeunt . . . Trinculo*]
CAPELL; *not in* F

282 **pickle** here alluding to his dousing in
'the foul lake', 4.1.183
283–4 **I shall . . . fly-blowing** i.e. I am
thoroughly preserved
287 **sirrah** The term expresses 'contempt,
reprimand, or assumption of authority on
the part of the speaker' (*OED*).
288 **sore** (a) sorry, inept (*OED* i. 8); (b) in
pain. The sense 'oppressive, severe' does
not seem relevant here.

290 **manners** both 'forms of behaviour' and
'moral character'
293 **trim** prepare, with implications of both
'make neat' and 'decorate'
295 **grace** forgiveness, favour
298 **luggage** the stolen garments; the term
is contemptuous, as in 4.1.231.
302 **waste** pass, occupy (with no pejorative
connotation: *OED* i. 8)

With such discourse as I not doubt shall make it
Go quick away: the story of my life,
And the particular accidents gone by
Since I came to this isle; and in the morn
I'll bring you to your ship, and so to Naples,
Where I have hope to see the nuptial
Of these our dear-belov'd solemnizèd,
And thence retire me to my Milan, where   310
Every third thought shall be my grave.

ALONSO   I long
To hear the story of your life, which must
Take the ear strangely.

PROSPERO   I'll deliver all,
And promise you calm seas, auspicious gales,
And sail so expeditious that shall catch
Your royal fleet far off. My Ariel, chick,
That is thy charge. Then to the elements
Be free, and fare thou well.—Please you draw near.

                                    Exeunt all

Epilogue *spoken by Prospero*
                Now my charms are all o'erthrown,
                And what strength I have's mine own,   320
                Which is most faint. Now 'tis true
                I must be here confined by you,
                Or sent to Naples. Let me not,

305 **accidents** events
307 **bring** accompany
309 **solemnizèd** accented on the second and
      fourth syllables
311 **Every . . . grave** This may be more than
      the conventional *memento mori*: see the
      Introduction, p. 55.
313 **Take** affect
      strangely wonderfully
      deliver relate
315 **sail . . . shall** a voyage so swift that it
      will; for the construction, see Abbott 279.
316 **far off** The other ships are a day's jour-
      ney ahead.
      chick an affectionate epithet; compare
      bird, 4.1.184.
318 **draw near** come in
319 **Epilogue** Prospero's epilogue is unique
      in the Shakespeare canon in that its

speaker declares himself not an actor in a
play but a character in a fiction. The
release he craves of the audience is the
freedom to continue his history beyond
the limits of the stage and the text. For the
octosyllabic couplets, compare the
epilogue to *A Midsummer Night's Dream*,
the Duke's choric moralizing in *Measure*
3.2.261–82, and Gower's speeches in
Pericles; the latter two examples also involve the same metrical roughness as
Prospero's epilogue.
322–7 Prospero puts himself in the position
      of Ariel, Caliban, Ferdinand, and the
      other shipwreck victims throughout the
      play, threatened with confinement,
      pleading for release from bondage; and
      his magical powers are now invested in
      the audience.

Since I have my dukedom got,
And pardoned the deceiver, dwell
In this bare island by your spell,
But release me from my bands
With the help of your good hands.
Gentle breath of yours my sails
Must fill, or else my project fails,   330
Which was to please. Now I want
Spirits to enforce, art to enchant;
And my ending is despair
Unless I be relieved by prayer,
Which pierces so that it assaults
Mercy itself, and frees all faults.
As you from crimes would pardoned be,
Let your indulgence set me free.   *Exit*

338 *In* F, *the cast of characters and location follow*

327 **bands** bonds
328 **your good hands** Sudden noises, and especially the clapping of hands, were thought to dissolve spells. Compare Prospero's call for silence at 4.1.126-7, 'or else our spell is marred'. The request for applause was the traditional epilogue of Roman comedy; and compare Puck's epilogue, *Dream* 5.1.444: 'Give me your hands if we be friends . . .'.
329 **Gentle breath** kind words (i.e. about the performance)
329-30 **breath . . . fill** This had been Ariel's charge at 314-17: the audience is now master and servant.
330 **project** See 5.1.1.
331 **want** lack
332 **enforce** control
333-4 **my ending . . . prayer** Warburton noted that 'this alludes to the old stories told of the despair of necromancers in their last moments, and the efficacy of the prayers of their friends for them' (cited in the Variorum). Kermode compares an exchange near the end of Marlowe's *Doctor*

Faustus: 'FAUSTUS. Ay, pray for me, pray for me . . . . SECOND SCHOLAR. Pray thou, and we will pray that God may have mercy upon thee', though he doubts the ultimate applicability of the analogy. However, see the Introduction, p. 56, n. I.
335 **pierces, assaults** In contrast to the 'gentle breath' required in l. 329, prayer is conceived as violent action. *Pierce* = 'touch or move deeply' (*OED* 5); *assault* = 'attack with reasoning or argument'; 'address with the object of persuading' (*OED* 3). Kermode compares Herbert's 'Prayer' (c. 1630), which calls it an 'Engine against th'Almighty'.
336 **Mercy** 'God's pitiful forbearance towards His creatures and forgiveness of their offences' (*OED* 1b), and hence a synecdoche for God.
frees frees from: see Abbott 200.
338 **indulgence** playing on the technical sense of remission of the punishment for sin

THE SEAMANSHIP OF ACT 1, SCENE 1

FROM A. F. Falconer, *Shakespeare and the Sea* (1964), pp. 37–9

The manœuvres described are difficult and some would be attempted only in an emergency.... The island is near, a violent onshore wind is blowing. The ship must weather or sail past the island, or else be driven so far in that running aground will be inevitable: 'fall to 't, yarely, or we run ourselves aground: bestir, bestir.' All that is done is meant to prevent this. To check the drift to leeward, the order, 'Take in the topsail', is given. The Boatswain knows that, above all else, they must have 'sea room' or room to manœuvre. If only they have that, the storm can do its worst, and so he shouts at it, 'Blow till thou burst thy wind, if room enough!' But still the ship makes toward the shore. The next order, 'Down with the top mast! yare! lower, lower!', is meant to ease the ship by reducing weight aloft, make the vessel roll less, and check the continuing drift shorewards.

Opinion was divided about striking topmasts, but there is a good reason for doing so here.

And as for the striking of the topmasts in this extremity of tempest, I am of his mind (though many are to the contrary) who holdeth that a ship is the wholesomer in the sea (though it be in a storm or tempest) when her topmasts are up, than when they are struck, and that she hath better way through it; so that when there is sea room enough it is the safest course not to strike them.[1]

Here, however, they do not have 'sea room enough' as the Boatswain's concern shows, and striking the topmast is justifiable.

To keep the ship close to the wind and away from the shore, until they can gain the open sea, is the aim of the next manœuvre. 'Bring her to try with main-course.' The main course is another term for the mainsail. 'Trying is to have no more sail forth but the mainsail, the tack aboard, the bowline set up, the sheet close aft, and the helm tied down close aboard.'[2] An important point in the manœuvre is that, 'A ship *a-try* with her mainsail (unless it be an extraordinary grown sea) will make her way two points afore the beam.'[3] That is what they hope the ship will do, but it does not happen. Instead of heading out to sea, it continues to be blown

---

[1] *Boteler's Dialogues*, ed. W. G. Perrin (Navy Records Society, Colchester, etc., 1929), p. 162.

[2] *The Life and Works of Sir Henry Mainwaring*, ed. G. E. Mainwaring and W. G. Perrin, 2 vols. (Navy Records Society, 1922), ii. 250.

[3] Ibid.

towards the island. In the hope of being able to keep clear of the leeshore, another order, 'Lay her a-hold!' is given. This is the only known example of the term in print, but it survives in New England, and is used there in handling yachts and sailing craft. It means to bring a vessel close to the wind so as to hold it or keep it.[1] To do so, more sail must be set and a further order follows, 'set her two courses'; that is, set the foresail in addition to the mainsail. The final directions, 'off to sea again'; 'lay her off', indicate what the result should be.

The ship is sound, the seamen are disciplined, the right orders are given. Some of the newer manœuvres of the day, even one that was debatable, have been tried, but all without success. . . .

Shakespeare could not have written a scene of this kind without taking great pains to grasp completely how a ship beset with these difficulties would have to be handled. He has not only worked out a series of manœuvres, but has made exact use of the professional language of seamanship. . . . He could not have come by this knowledge from books, for there were no works on seamanship in his day, nor were there any nautical word lists or glossaries, though several manuals had been published on navigation which, however, is a different art.

---

[1] The derivation of this term and its connection with 'haul' are discussed in an article in the *American Neptune*, 5 (July 1951), pp. 29–14.

THE STRACHEY LETTER

From *Purchas his Pilgrimes* (1625), Part 4, Book 9, Chapter 6
(pp. 1734 ff.)

A true repertory of the wreck and redemption of Sir Thomas Gates, Knight, upon and from the islands of the Bermudas, his coming to Virginia, and the estate of the colony then and after under the government of the Lord La Warre. July 15, 1610, written by William Strachey, Esquire.

I

A most dreadful tempest (the manifold deaths whereof are here to the life described),[1] their wreck on Bermuda, and the description of those islands.

. . . We were within seven or eight days at the most . . . of making Cape Henry upon the coast of Virginia when on St James his day, July 14, being Monday (preparing for no less all the black night before), the clouds gathering thick upon us, and the winds singing and whistling most unusually, which made us to cast off our pinnace (towing the same until then astern), a dreadful storm and hideous began to blow from out the north-east, which swelling and roaring as it were by fits, some hours with more violence than others, at length did beat all light from heaven, which like an hell of darkness turned black upon us, so much the more fuller of horror, as in such cases horror and fear use to overrun the troubled and overmastered senses of all, which, taken up with amazement, the ears lay so sensible to the terrible cries and murmurs of the winds and distraction of our company, as who was most armed and best prepared was not a little shaken. For surely . . . as death comes not so sudden nor apparent, so he comes not so elvish and painful (to men especially even then in health and perfect habitudes of body) as at sea; who comes at no time so welcome, but our frailty (so weak is the hold of hope in miserable demonstrations of danger) it makes guilty of many contrary changes and conflicts. For indeed, death is accompanied at no time nor place with circumstances everyway so uncapable of particularities of goodness and inward comforts as at sea. . . .

For four and twenty hours the storm in a restless tumult had blown so exceedingly as we could not apprehend in our imaginations any possibility

[1] In fact, Strachey reports no deaths as a result of the tempest.

of greater violence; yet did we still find it not only more terrible but more constant, fury added to fury, and one storm urging a second more outrageous than the former, whether it so wrought upon our fears, or indeed met with new forces. Sometimes strikes in our ship amongst women and passengers, not used to such hurly and discomforts, made us look one upon the other with troubled hearts and panting bosoms, our clamours drowned in the winds, and the winds in thunder. Prayers might well be in the heart and lips, but drowned in the outcries of the officers; nothing heard that could give comfort, nothing seen that might encourage hope. . . . Our sails wound up lay without their use, and if at any time we bore but a hullock,[1] or half forecourse, to guide her before the sea, six and sometimes eight men were not enough to hold the whipstaff in the steerage and the tiller below in the gunner room, by which may be imagined the strength of the storm, in which the sea swelled above the clouds and gave battle to the heaven. It could not be said to rain; the waters like whole rivers did flood in the air. And this I did still observe, that whereas upon the land, when a storm hath poured itself forth once in drifts of rain, the wind as beaten down and vanquished therewith not long after endureth, here the glut of water, as if throttling the wind erewhile, was no sooner a little emptied and qualified but instantly the winds, as having gotten their mouths now free and at liberty, spake more loud and grew more tumultuous and malignant. What shall I say? Winds and seas were as mad as fury and rage could make them. For mine own part, I had been in some storms before, . . . yet all that I had ever suffered gathered together might not hold in comparison with this: there was not a moment in which the sudden splitting or instant oversetting of the ship was not expected.

Howbeit this was not all; it pleased God to bring a greater affliction yet upon us, for in the beginning of the storm we had received likewise a mighty leak. . . .

Our governor, upon the Tuesday morning (at what time, by such who had been below in the hold, the leak was first discovered), had caused the whole company, about one hundred and forty, besides women, to be equally divided into three parts, and opening the ship in three places (under the forecastle, in the waist, and hard by the bittacle[2]) appointed each man where to attend; and thereunto every man came duly upon his watch, took the bucket or pump for one hour, and rested another. Then men might be seen to labour, I may well say, for life, and the better sort, even our governor and admiral themselves, not refusing their turn and to spell each the other, to give example to other. The common sort stripped naked, as men in galleys, the easier both to hold out and to shrink from

---

[1] 'A small part of a sail let out in a gale to keep the ship's head to the sea' (*OED*).
[2] Or binnacle, 'a box on the deck of a ship near the helm, in which the compass is placed' (*OED*).

under the salt water, which continually leapt in among them, kept their eyes waking and their thoughts and hands working, with tired bodies and wasted spirits, three days and four nights destitute of outward comfort, and desperate of any deliverance, testifying how mutually willing they were yet by labour to keep each other from drowning, albeit each one drowned whilst he laboured.

Once so huge a sea broke upon the poop and quarter upon us as it covered our ship from stern to stem like a garment or a vast cloud; it filled her brim-full for a while within, from the hatches up to the spar-deck. This source or confluence of water was so violent as it rushed and carried the helmsman from the helm and wrested the whipstaff out of his hand, which so flew from side to side that when he would have ceased the same again, it so tossed him from starboard to larboard as it was God's mercy it had not split him. It so beat him from his hold and so bruised him as a fresh man hazarding in by chance fell fair with it, and by main strength bearing somewhat up made good his place, and with much clamour encouraged and called upon others, who gave her now up, rent in pieces and absolutely lost. Our governor was at this time below at the capstan, both by his speech and authority heartening every man unto his labour. It struck him from the place where he sat, and grovelled him and all us about him on our faces, beating together with our breaths all thoughts from our bosoms else than that we were now sinking. For my part, I thought her already in the bottom of the sea; and I have heard him say wading out of the flood thereof, all his ambition was but to climb up above hatches to die *in aperto coelo*, and in the company of his old friends. . . .

During all this time the heavens looked so black upon us that it was not possible the elevation of the pole might be observed, nor a star by night nor sunbeam by day was to be seen. Only upon the Thursday night Sir George Summers being upon the watch had an apparition of a little round light, like a faint star, trembling and streaming along with a sparkling blaze half the height upon the mainmast, and shooting sometimes from shroud to shroud, 'tempting to settle as it were upon any of the four shrouds; and for three or four hours together, or rather more, half the night it kept with us, running sometimes along the mainyard to the very end and then returning. At which Sir George Summers called divers about him and showed them the same, who observed it with much wonder and carefulness; but upon a sudden, towards the morning watch, they lost the sight of it, and knew not what way it made. The superstitious seamen make many constructions of this sea-fire, which nevertheless is usual in storms: the same, it may be, which the Grecians were wont in the Mediterranean to call Castor and Pollux, of which if only one appeared without the other, they took it for an evil sign of great tempest. The Italians and such who lie open to the Adriatic and Tyrrhene Sea call it a sacred body, *corpo sancto*; the

Spaniards call it Saint Elmo, and have an authentic and miraculous legend for it. Be it what it will, we laid other foundations of safety or ruin than in the rising or falling of it; could it have served us now miraculously to have taken our height by, it might have strucken amazement and a reverence in our devotions, according to the due of a miracle. But it did not light us any whit the more to our known way, who ran now (as do hoodwinked men) at all adventures, sometimes north and north-east, then north and by west, and in an instant again varying two or three points, and sometimes half the compass. East and by south we steered away as much as we could to bear upright, which was no small carefulness nor pain to do, albeit we much unrigged our ship, threw overboard much luggage, many a trunk and chest (in which I suffered no mean loss) and staved many a butt of beer, hogsheads of oil, cider, wine, and vinegar, and heaved away all our ordnance on the starboard side, and had now purposed to have cut down the mainmast the more to lighten her; for we were much spent, and our men so weary as their strengths together failed them with their hearts, having travailed now from Tuesday till Friday morning, day and night, without either sleep or food; for the leakage taking up all the hold, we could neither come by beer nor fresh water; fire we could keep none in the cook-room to dress any meat; and carefulness, grief, and our turn at the pump or bucket were sufficient to hold sleep from our eyes . . . and from Tuesday noon till Friday noon we bailed and pumped two thousand ton, and yet do what we could, when our ship held least in her (after Tuesday night second watch) she bore ten foot deep, at which stay our extreme working had kept her one eight glasses,[1] forbearance whereof had instantly sunk us, and it being now Friday, the fourth morning, it wanted little but that there had been a general determination to have shut up hatches and, commending our sinful souls to God, committed the ship to the mercy of the sea; surely that night we must have done it, and that night had we then perished. But see the goodness and sweet introduction of better hope by our merciful God given unto us. Sir George Summers, when no man dreamed of such happiness, had discovered and cried land. Indeed, the morning now three-quarters spent had won a little clearness from the days before, and it being better surveyed, the very trees were seen to move with the wind upon the shore side; whereupon our governor commanded the helmsman to bear up, the boatswain sounding at the first found it thirteen fathom, and when we stood a little in, seven fathom, and presently heaving his lead the third time had ground at four fathom; and by this we had got her within a mile under the south-east point of the land, where we had somewhat smooth water. But having no hope to save her by coming to an anchor in the same, we were enforced to run her ashore

---

[1] i.e. held the leakage at the same level for four hours (the glasses are mariners' half-hour measures).

as near the land as we could, which brought us within three-quarters of a mile of shore, and by the mercy of God unto us, making out our boats, we had ere night brought all our men, women, and children, about the number of one hundred and fifty, safe into the island.

We found it to be the dangerous and dreaded island, or rather islands, of the Bermuda, whereof let me give your ladyship a brief description before I proceed to my narration. And that the rather, because they be so terrible to all that ever touched on them, and such tempests, thunders, and other fearful objects are seen and heard about them that they may be called commonly the Devil's Islands, and are feared and avoided of all sea travellers alive, above any other place in the world. Yet it pleased our merciful God to make even this hideous and hated place both the place of our safety and means of our deliverance.

And hereby also I hope to deliver the world from a foul and general error, it being counted of most that they can be no habitation for men, but rather given over to devils and wicked spirits; whereas indeed we find them now by experience to be as habitable and commodious as most countries of the same climate and situation, insomuch as if the entrance into them were as easy as the place itself is contenting, it had long ere this been inhabited as well as other islands. Thus shall we make it appear that Truth is the daughter of Time, and that men ought not to deny everything which is not subject to their own sense.

The Bermudas be broken islands, five hundred of them in manner of an archipelagus (at least if you may call them all islands that lie, how little soever, into the sea, and by themselves), of small compass, some larger yet than other, as time and the sea hath won from them and eaten his passage through, and all now lying in the figure of a croissant within the circuit of six or seven leagues at the most, albeit at first it is said of them that they were thirteen or fourteen leagues, and more in longitude, as I have heard.[1]

These islands are often afflicted and rent with tempests, great strokes of thunder, lightning, and rain in the extremity of violence. . . .

The soil of the whole island is one and the same, the mould dark, red, sandy, dry, and uncapable, I believe, of any of our commodities or fruits. Sir George Summers in the beginning of August squared out a garden . . . and sowed musk melons, peas, onions, radish, lettuce, and many English seeds and kitchen herbs. All which in some ten days did appear above ground, but whether by the small birds, of which there were many kinds, or by flies (worms I never saw any, nor any venomous thing, as toad or snake or any creeping beast hurtful, only some spiders, which as many

---

[1] Strachey goes on to cite 'the testimony of Gonzalus Ferdinandus Oviedus', author of a history of the West Indies, to the effect that the islands seem to have diminished since their discovery. Bullough (p. 281) suggests this as a source for the names of Gonzalo and Ferdinand.

affirm are signs of great store of gold; but they were long and slender-leg spiders, and whether venomous or no I know not—I believe not, since we should still find them amongst our linen in our chests and drinking cans, but we never received any danger from them; a kind of melolontha, or black beetle, there was, which bruised, gave a savour like many sweet and strong gums pounded together)—whether, I say, hindered by these or by the condition or vice of the soil, they came to no proof, nor thrived. It is like enough that the commodities of the other western islands would prosper there, as vines, lemons, oranges, and sugar canes: our governor made trial of the latter and buried some two or three in the garden mould, which were reserved in the wreck amongst many which we carried to plant here in Virginia, and they began to grow; but the hogs breaking in both rooted them up and ate them. There is not through the whole islands either champaign ground, valleys, or fresh rivers. They are full of shaws[1] of goodly cedar, fairer than ours here of Virginia, the berries whereof our men seething, straining, and letting stand some three or four days made a kind of pleasant drink. . . .

Likewise there grow great store of palm trees, not the right Indian palms, such as . . . are called cocos, . . . nor of those kind of palms which bears dates, but a kind of cimarrons[2] or wild palms in growth, fashion, leaves and branches, resembling those true palms; for the tree is high and straight, sappy and spongeous, unfirm for any use; no branches but in the uppermost part thereof, and in the top grow leaves about the head of it (the most inmost part whereof they call palmeto, and it is the heart and pith of the same trunk, so white and thin as it will peel off into pleats as smooth and delicate as white satin, into twenty folds, in which a man may write as in paper) where they spread and fall downward about the tree like an overblown rose or saffron flower not early gathered; so broad are the leaves as an Italian umbrella: a man may well defend his whole body under one of them from the greatest storm rain that falls. . . .

Sure it is that there are no rivers nor running springs of fresh water to be found upon any of them. When we came first we digged and found certain gushings and soft bubblings, which being either in bottoms or on the side of hanging ground were only fed with rain water, which nevertheless soon sinketh into the earth and vanisheth away, or emptieth itself out of sight into the sea, without any channel above or upon the superficies of the earth; for according as their rains fell, we had our wells and pits (which we digged) either half full or absolute exhausted and dry, howbeit some low bottoms (which the continual descent from the hills filled full, and in those flats could have no passage away) we found to continue as

[1] Thickets.
[2] Escaped slaves who lived in the hills and forests, hence wild; the word is Spanish, and is not recorded in the *OED*.

fishing ponds or standing pools continually, summer and winter, full of fresh water.

The shore and bays round about when we landed first afforded great store of fish, and that of divers kinds, and good. . . . We have taken also from under the broken rocks crayfish often greater than any of our best English lobsters, and likewise abundance of crabs, oysters, and whelks. . . .

Fowl there is in great store, small birds, sparrows fat and plump like a bunting, bigger than ours, robins of divers colours, green and yellow, ordinary and familiar in our cabins, and other of less sort. White and grey hernshaws, bitterns, teal, snipes, crows, and hawks, of which in March we found divers aeries, goshawks, and tercels; oxen-birds, cormorants, bald coots, moorhens, owls, and bats in great store. . . . A kind of web-footed fowl there is, of the bigness of an English green plover, or sea-mew,[1] which all the summer we saw not, and in the darkest nights of November and December (for in the night they only feed) they would come forth, but not fly far from home, and hovering in the air and over the sea made a strange hollow and harsh howling. . . . Our men found a pretty way to take them, which was by standing on the rocks or sands by the seaside and holloing, laughing and making the strangest outcry that possibly they could, with the noise whereof the birds would come flocking to that place and settle upon the very arms and head of him that so cried, and still creep nearer and nearer, answering the noise themselves, by which our men would weigh them with their hand, and which weighed heaviest they took for the best and let the others alone, and so our men would take twenty dozen in two hours of the chiefest of them; and they were a good and well-relished fowl, fat and full as a partridge. . . .

We had knowledge that there were wild hogs upon the island, at first by our own swine preserved from the wreck and brought to shore; for they straying into the woods, an huge wild boar followed down to our quarter, which at night was watched and taken . . . and there be thousands of them in the islands, and at that time of the year, in August, September, October, and November they were well fed with berries that dropped from the cedars and the palms, and in our quarter we made sties for them. . . .

The tortoise is reasonable toothsome (some say) wholesome meat. I am sure our company liked the meat of them very well, and one tortoise would go further amongst them than three hogs. One turtle (for so we called them) feasted well a dozen messes, appointing six to every mess. It is such a kind of meat as a man can neither absolutely call fish nor flesh, keeping most what in the water, and feeding upon sea grass like a heifer in the

---

[1] Strachey's text reads 'sea-meawe'; Bullough believes that this probably explains Caliban's 'scamels'. The suggestion would be more persuasive if Strachey had used the alternate form 'sea-mell'.

bottom of the coves and bays, and laying their eggs (of which we should find five hundred at a time in the opening of a she turtle) in the sand by the shore side, and so covering them close leave them to the hatching of the sun. . . . Their eggs are as big as geese eggs, and themselves grown to perfection, bigger than great round targets.

II

[The settlers fall into dissension. A conspiracy is discovered and aborted.]

[P. 1744] In these dangers and devilish disquiets (whilst the almighty God wrought for us and sent us, miraculously delivered from the calamities of the sea, all blessings upon the shore to content and bind us to gratefulness) thus enraged amongst ourselves to the destruction of each other, into what a mischief and misery had we been given up had we not had a governor with his authority to have suppressed the same? Yet was there a worse practice, faction and conjuration afoot, deadly and bloody, in which the life of our governor, with many others, were threatened, and could not but miscarry in his fall. But such is ever the will of God (who in the execution of his judgements breaketh the firebrands upon the head of him who first kindleth them), there were who conceived that our governor indeed neither durst nor had authority to put in execution or pass the act of justice upon anyone, how treacherous or impious soever; their own opinions so much deceiving them for the unlawfulness of any act which they would execute, daring to justify among themselves that if they should be apprehended before the performance, they should happily suffer as martyrs. They persevered therefore not only to draw unto them such a number and associates[1] as they could work in to the abandoning of our governor and to the inhabiting of this island. They had now purposed to have made a surprise of the storehouse and to have forced from thence what was therein, either of meal, cloth, cables, arms, sails, oars, or what else it pleased God that we had recovered from the wreck and was to serve our general necessity and use, either for the relief of us while we stayed here, or for the carrying of us from this place again when our pinnace should have been furnished.

But as all giddy and lawless attempts have always something of imperfection, and that as well by the property of the action, which holdeth of disobedience and rebellion (both full of fear), as through the ignorance of the devisers themselves, so in this, besides those defects, there were some of the association who, not strong enough fortified in their own conceits, broke from the plot itself, and before the time was ripe for the execution thereof discovered the whole order and every agent and actor thereof, who

---

[1] i.e. so many and such associates.

nevertheless were not suddenly apprehended, by reason the confederates were divided and separated in place, some with us, and the chief with Sir George Summers in his island (and indeed all his whole company); but good watch passed upon them, every man from thenceforth commanded to wear his weapon, without which, before, we freely walked from quarter to quarter and conversed among ourselves; and every man advised to stand upon his guard, his own life not being in safety, whilst his next neighbour was not to be trusted. The sentinels and nightwarders doubled, the passages of both the quarters were carefully observed, by which means nothing was further attempted until a gentleman amongst them, one Henry Paine, the thirteenth of March, full of mischief, and every hour preparing something or other, stealing swords, adzes, axes, hatchets, saws, augurs, planes, mallets, etc. to make good his own bad end, his watch night coming about, and being called by the captain of the same to be upon the guard, did not only give his said commander evil language, but struck at him, doubled his blows, and when he was not suffered to close with him, went off the guard, scoffing at the double diligence and attendance of the watch appointed by the governor for much purpose, as he said; upon which the watch telling him if the governor should understand of his insolency, it might turn him to much blame, and haply be as much as his life were worth. The said Paine replied with a settled and bitter violence and in such unreverent terms as I should offend the modest ear too much to express it in his own phrase, but the contents were how the governor had no authority of that quality to justify upon anyone how mean soever in the colony an action of that nature, and therefore let the governor (said he) kiss etc. Which words being with the omitted additions brought the next day unto every common and public discourse, at length they were delivered over to the governor, who, . . . calling the said Paine before him, and the whole company, where (being soon convinced both by the witness of the commander and many which were upon the watch with him) our governor, who had now the eyes of the whole colony fixed upon him, condemned him to be instantly hanged; and the ladder being ready, after he had made many confessions, he earnestly desired, being a gentleman, that he might be shot to death; and towards the evening he had his desire, the sun and his life setting together.

But for the other which were with Sir George, upon the Sunday following . . . by a mutual consent forsook their labour and Sir George Summers, and like outlaws betook them to the wild woods. Whether mere rage, and greediness after some little pearl (as it was thought) wherewith they conceived they should forever enrich themselves, and saw how to obtain the same easily in this place, or whether the desire forever to inhabit here, or whatever other secret else moved them thereunto, true it is, they sent an audacious and formal petition to our governor subscribed with all their names and seals, not only entreating him that they might stay here, but with great art importuned him that he would perform other conditions with them, and not wave nor evade from some of his own promises, as namely to furnish each of them with two suits of apparel and contribute meal rateably for one whole year, so much among them as they had weekly now, which was one pound and an half a week, for such had been our proportion for nine months. Our governor answered this their petition, writing to Sir George Summers to this effect.

[The petition is granted, but only two of the malcontents choose to remain in Bermuda. The rest of the company set sail for Virginia.]

III

[When they reach Jamestown, they find it 'full of misery and misgovernment', lacking the presence of the governor. A set of laws and regulations is quickly promulgated. Strachey describes the fertility of the land.]

[P. 1750] What England may boast of, having the fair hand of husbandry to manure and dress it, God and nature have favourably bestowed upon this country, and as it hath given unto it both by situation, height, and soil all those (past hopes) assurances which follow our well-planted native country, and others lying under the same influence, if, as ours, the country and soil might be improved and drawn forth, so hath it endowed it, as is most certain, with many more, which England fetcheth far unto her from elsewhere. For first we have experience, and even our eyes witness (how young soever we are to the country) that no country yieldeth goodlier corn nor more manifold increase; large fields we have, as prospects of the same, and not far from our palisado. Besides, we have thousands of goodly vines in every hedge and bosk running along the ground, which yield a plentiful grape in their kind. Let me appeal then to knowledge, if these natural vines were planted, dressed, and ordered by skilful vignerons whether we might not make a perfect grape and fruitful vintage in short time? And we have made trial of our own English seeds, kitchen herbs, and roots, and find them to prosper as speedily as in England.

IV

[In the course of describing the settlers' troubles with the Indians, Strachey recounts how one of Gates's men attempts to recover a longboat that has been stranded near a native encampment.]

[P. 1755] . . . certain Indians, watching the occasion, seized the poor fellow and led him up into the woods and sacrificed him. It did not a little trouble the lieutenant-governor, who since his first landing in the country, how justly soever provoked, would not by any means be wrought to a violent proceeding against them for all the practices of villainy with which they daily endangered our men, thinking it possible by a more tractable course to win them to a better condition; but now, being startled by this, he well perceived how little a fair and noble entreaty works upon a barbarous disposition, and therefore in some measure purposed to be revenged.[1]

[Strachey closes by quoting a portion of the Virginia Company's pamphlet *A True Declaration of Virginia*.]

[P. 1757] The ground of all those miseries was the permissive providence of God, who, in the forementioned violent storm separated the head from the body, all the vital powers of regiment being exiled with Sir Thomas Gates in those infortunate (yet fortunate) islands. The broken remainder of those supplies made a greater shipwreck in the continent of Virginia by the tempest of dissension; every man overvaluing his own worth would be a commander, every man underprizing another's value denied to be commanded.

The next fountain of woes was secure negligence and improvidence, when every man sharked for his present booty, but was altogether careless of succeeding penury. Now I demand whether Sicilia or Sardinia, sometimes the barns of Rome, could hope for increase without manuring? A colony is therefore denominated because they should be *coloni*, the tillers of the earth and stewards of fertility; our mutinous loiterers would not sow with providence, and therefore they reaped the fruits of too dear-bought repentance. An incredible example of their idleness is the report of Sir Thomas Gates, who affirmeth that after his first coming thither he hath seen some of them eat their fish raw, rather than they would go a stone's cast to fetch wood and dress it. *Dii laboribus omnia vendunt*, God sells us all things for our labour, when Adam himself might not live in paradise without dressing the garden.

[1] Purchas's marginal note on this incident reads, 'Can a leopard change his spots? Can a savage remaining a savage be civil? Were not we ourselves made and not born civil in our progenitors' days? The Roman swords were best teachers of civility to this and other countries near us.'

THE MUSIC

No Shakespeare play calls for more music, and of more various kinds, than *The Tempest*. The following summary indicates its range and pervasiveness:

1.2.373.2 Ariel enters *playing and singing* the song 'Come unto these yellow sands', followed by 'Full fathom five thy father lies'. Both have refrains of other voices.

2.1.182.1 Ariel enters *playing solemn music*.

    294.2 Ariel enters *with music and song* and sings 'While you here do snoring lie'.

2.2. 40.2 Stephano enters *singing* 'I shall no more to sea, to sea' and 'The master, the swabber, the boatswain, and I'.

    172 Caliban *sings drunkenly* 'Farewell, master, farewell, farewell', and then 'No more dams I'll make for fish'.

3.2.119.1 Stephano, Trinculo and Caliban *sing* 'Flout 'em and cout 'em' to the wrong tune; Ariel supplies the right one on a tabor and pipe.

3.3. 17.1 *Solemn and strange music* accompanies the appearance of *several strange shapes bringing in a banquet*.

    82.1 *Soft music* accompanies the reappearance and dance of the *shapes*.

4.1. 58.1 *Soft music* introduces Prospero's masque.

    106.2 Juno and Ceres *sing* 'Honour, riches, marriage-blessing', presumably with instrumental accompaniment.

    138 Reapers perform *a graceful dance*, interrupted by *a strange hollow and confused noise*.

5.1. 52 Prospero invokes 'some heavenly music' as an 'airy charm' to restore the senses of Alonso and the other Neapolitans.

    87.1 Ariel *sings* 'Where the bee sucks, there suck I'.

The only part of the original music that can with any confidence be said to survive is the settings of two of Ariel's songs, 'Full fathom five thy father lies' in Act 1, Scene 2 and 'Where the bee sucks, there suck I' in Act 5, Scene 1. These are reprinted on pp. 223–6. They were first published in John Wilson's *Cheerful Ayres*, 1660 [for 1659], where they are attributed to Robert Johnson. The songs also appear in several manuscripts.[1] A third piece has sometimes been included as part of the original music for the play. This is a dance entitled 'The Tempest' in British Library Add. MS 10444, a miscellaneous collection of masque music. The suggestion that it belongs to Shakespeare's play was made, very tentatively, by W. J. Lawrence,[2] and it is included, equally tentatively, by Kermode in the Arden edition. J. M. Nosworthy, however, following John P. Cutts, believes that it 'clearly belongs to the play'.[3] But as Andrew Sabol points out, the title itself is an argument against the attribution: 'all the items in this part of Brit. Lib. Add. 10444 are dances, not background music, and their headings, when descriptive, characterize the dance or dancers rather than reflect the titles of masques or plays.'[4] That is, the titles of pieces in this manuscript identify the name or character of the particular dance, not the work in which it was performed. There is no 'tempest dance' in *The Tempest*.

Robert Johnson (*c.*1582–1633) was the son of John Johnson, one of Queen Elizabeth I's lutenists. In 1596 he entered the service of Sir George Carey, who in the same year succeeded to his father's title of Baron Hunsdon, and in the year following to his father's post of Lord Chamberlain, an office he held till his death in 1603. Hunsdon was, of course, the patron of Shakespeare's company, and it was perhaps from this that Johnson's later association with the King's Men derived. In 1604 Johnson became lutenist to James I, and remained in the royal employ until his death. He was also appointed musician to Prince Henry, and to Charles as Prince of Wales.

Johnson composed instrumental music for Ben Jonson's *Oberon* in 1611, and for both Chapman's and Beaumont's masques for Princess Elizabeth's wedding in 1613; he may have contributed to court masques earlier, and probably continued to do so sporadically until about 1621, when he set 'From the famous Peak of Derby' for Jonson's *Gypsies Metamorphosed*. He was employed 'particularly', Ian Spink writes, 'when supernatural or otherwise bizarre dances were required, as they often were'.[5] In the same period he frequently provided vocal music for the

[1] The songs are in the Birmingham City Reference Library MS 57316, pp. 87–8, the Folger Shakespeare Library MS 747.1, fols. 9v–13v, and in three manuscript partbooks from the Filmer Collection at Yale, shelf number Ma21.F48.A13(a–c). 'Where the bee sucks' also appears in the Bodleian Library ms Don.c.57, p. 74, where the song is ascribed to Wilson, from whose text it presumably derives.

[2] *Music and Letters*, 3 (1922), pp. 49–58.

[3] 'Music and its Function in Shakespeare's Romances', *Shakespeare Survey* 11 (Cambridge, 1958), pp. 60–9. Cutts's edition of the *Tempest* music is in *La Musique de scène de la troupe de Shakespeare* (Paris 1959).

[4] *Four Hundred Songs and Dances from the Stuart Masque* (Providence, 1978), p. 576.

[5] *Robert Johnson : Ayres, Songs and Dialogues* (*The English Lute-Songs*, 2nd Series, vol. 17) (1961), p. iii.

King's Men. Songs by him survive for a number of Beaumont and Fletcher plays, for Webster's *Duchess of Malfi*, Jonson's *The Devil is an Ass* ('Have you seen but a bright lily grow'), and Middleton's *The Witch* ('Come away, Hecate', which also appears in the surviving version of *Macbeth*). He may probably be credited as well with 'Hark, hark the lark' from *Cymbeline* and 'Get you hence, for I must go' from *The Winter's Tale*.

The standard edition of Johnson's *Ayres* is that of Ian Spink, whose transcriptions are used here. There has been some question about whether the settings John Wilson printed were indeed those composed for the play, since 'Full fathom five' omits the burden 'ding dong'. But we have no way of knowing what sort of manuscript Wilson was transcribing, or what editorial liberties he may have taken in turning a dramatic song into an ayre for private performance. The identification must stop short of absolute certainty; this is as close as we are likely to come to the music for the play in Shakespeare's time.[1]

---

[1] The most helpful and reliable accounts of *The Tempest* music are in Peter J. Seng, *The Vocal Songs in the Plays of Shakespeare* (Cambridge, Mass., 1967), pp. 251–65, and John Stevens, 'Shakespeare and the Music of the Elizabethan Stage', in *Shakespeare in Music*, ed. Phyllis Hartnoll (1964), pp. 45–7. David Lindley gives a detailed and original analysis of the dramatic function of music in the play in 'Music, masque and meaning in *The Tempest*', in *The Court Masque*, ed. David Lindley (Manchester, 1984), pp. 47–59. The article on Shakespeare by F. W. Sternfeld and Eric Walter White in *The New Grove Dictionary of Music and Musicians* (1980), xvii 218, includes an extensive bibliography.

FULL FATHOM FIVE

S. & B. 5513

From *Robert Johnson, Ayres, Songs and Dialogues*, ed. Ian Spink (Stainer and Bell, 1961), pp. 24–7. © Stainer & Bell Ltd. Reprinted by permission. Sole American agent, Galaxy Music Corporation, New York.

Sea nymphs hour-ly ring his knell, Hark now I hear them, hark now I hear them, Ding, dong, bell. Ding, dong, ding, dong, bell; Ding, dong, ding, dong, Ding, dong, ding, dong, bell.

WHERE THE BEE SUCKS

S. & B. 5513

S.& B. 5513

FLORIO'S MONTAIGNE

FROM Michel de Montaigne, *The Essays*, translated by John Florio (1603)

Book 1, Chapter 30 (pp. 100–7) 'Of the Cannibals'
At what time King Pyrrhus came into Italy, after he had surveyed the marshalling of the army which the Romans sent against him, 'I wot not', said he, 'what barbarous men these are' (for so were the Grecians wont to call all strange nations), 'but the disposition of the army which I see is nothing barbarous.' So said the Grecians of that which Flaminius sent into their country, and Philip, viewing from a tower the order and distribution of the Roman camp in his kingdom under Publius Sulpitius Galba. Lo how a man ought to take heed lest he overweeningly follow vulgar opinions, which should be measured by the rule of reason and not by the common report. I have had long time dwelling with me a man who for the space of ten or twelve years had dwelt in that other world which in our age was lately discovered in those parts where Villegaignon first landed and sur-named Antarctic France.[1] This discovery of so infinite and vast a country seemeth worthy great consideration. I wot not whether I can warrant myself that some other be not discovered hereafter, sithence so many worthy men, and better learned than we are, have so many ages been deceived in this. I fear me our eyes be greater than our bellies, and that we have more curiosity than capacity. We embrace all, but we fasten nothing but wind. Plato maketh Solon to report that he had learnt of the priests of the city of Saïs in Egypt that whilom, and before the general deluge, there was a great island called Atlantides situated at the mouth of the Strait of Gibraltar, which contained more firm land than Afric and Asia together. And that the kings of that country, who did not only possess that island but had so far entered into the mainland that of the breadth of Afric they held as far as Egypt and of Europe's length as far as Tuscany, and that they undertook to invade Asia and to subdue all the nations that compass the Mediterranean Sea to the gulf of Mare Maggiore,[2] and to that end they traversed all Spain, France, and Italy, so far as Greece, where the Athenians made head against them; but that a while after, both the Athenians themselves and that great island were swallowed up by the deluge. It is very likely this extreme ruin of waters wrought strange altera-tions in the habitations of the earth, as some hold that the sea hath divided Sicily from Italy—

[1] The landing was in Brazil in 1557.
[2] The Black Sea.

Haec loca vi quondam, et vasta convulsa ruina
Dissiluisse ferunt, cum protinus utraque tellus
Una foret[1]
Men say sometimes this land by that forsaken,
And that by this, were split and ruin-shaken,
Whereas till then both lands as one were taken—

Cyprus from Syria, the island of Negropont[2] from the mainland of Boeotia, and in other places joined lands that were sundered by the sea, filling with mud and sand the channels between them.

. . . sterilisque diu palus aptaque remis
Vicinas urbes alit, et grave sentit aratrum.[3]
The fen long barren, to be rowed in, now
Both feeds the neighbour towns and feels the plough.

But there is no great apparance[4] the said island should be the new world we have lately discovered, for it well nigh touched Spain, and it were an incredible effect of inundation to have removed the same more than twelve hundred leagues, as we see it is. Besides, our modern navigations have now almost discovered that it is not an island but rather firm land, and a continent, with the East Indies on one side and the countries lying under the two poles on the other, from which if it be divided, it is with so narrow a strait and interval that it no way deserveth to be named an island; for it seemeth there are certain motions in these vast bodies, some natural and other some febricitant,[5] as well as in ours. When I consider the impression my river of Dordogne worketh in my time toward the right shore of her descent, and how much it hath gained in twenty years, and how many foundations of divers houses it hath overwhelmed and violently carried away, I confess it to be an extraordinary agitation; for should it always keep one course, or had it ever kept the same, the figure of the world had ere this been overthrown; but they are subject to changes and alterations. Sometimes they overflow and spread themselves on one side, sometimes on another, and other times they contain themselves in their natural beds or channels. I speak not of sudden inundations, whereof we now treat the causes. In Médoc alongst the sea coast my brother the Lord of Arsac may see a town of his buried under the sands which the sea casteth up before it; the tops of some buildings are yet to be discerned. His rents and demesnes have been changed into barren pastures. The inhabitants thereabouts affirm that some years since, the sea encroacheth so much upon them that they have lost four leagues of firm land: these sands are her forerunners. And we see great hillocks of gravel moving, which march half a league

---

[1] Virgil, *Aeneid*, iii. 414–16.
[2] Euboea.
[3] Horace, *Ars Poetica*, 65–6.
[4] Likelihood.
[5] Feverish, caused by disease.

before it and usurp on the firm land. The other testimony of antiquity to which some will refer this discovery is in Aristotle (if at least that little book of unheard of wonders be his), where he reporteth that certain Carthaginians having sailed athwart the Atlantic Sea without the Strait of Gibraltar, after a long time they at last discovered a great fertile island replenished with goodly woods and watered with great and deep rivers far distant from all land, and that both they and others, allured by the goodness and fertility of the soil, went thither with their wives, children, and household, and there began to habituate and settle themselves. The lords of Carthage, seeing their country by little and little to be dispeopled, made a law and express inhibition that upon pain of death no more men should go thither, and banished all that were gone thither to dwell, fearing (as they said) that in success of time they would so multiply as they might one day supplant them and overthrow their own estate. This narration of Aristotle hath no reference unto our newfound countries. This servant I had was a simple and rough-hewn fellow, a condition fit to yield a true testimony. For subtle people may indeed mark more curiously and observe things more exactly, but they amplify and gloss them, and the better to persuade and make their interpretations of more validity, they cannot choose but somewhat alter the story. They never represent things truly, but fashion and mask them according to the visage they saw them in, and to purchase credit to their judgement and draw you on to believe them they commonly adorn, enlarge, yea, and hyperbolize the matter. Wherein is required either a most sincere reporter or a man so simple that he may have no invention to build upon and to give a true likelihood unto false devices, and be not wedded to his own will. Such a one was my man, who besides his own report hath many times showed me divers mariners and merchants whom he had known in that voyage. So am I pleased with his information that I never enquire what cosmographers say of it. We had need of topographers to make us particular narrations of the places they have been in. For some of them, if they have the advantage of us that they have seen Palestine, will challenge a privilege to tell us news of all the world besides. I would have every man write what he knows and no more —not only in that, but in all other subjects. For one may have particular knowledge of the nature of one river and experience of the quality of one fountain that in other things knows no more than another man, who nevertheless to publish this little scantling will undertake to write of all the physics. From which vice proceed divers great inconveniences. Now (to return to my purpose) I find, as far as I have been informed, there is nothing in that nation that is either barbarous or savage, unless men call that barbarism which is not common to them. As indeed we have no other aim of truth and reason than the example and idea of the opinions and customs of the country we live in. Where is ever perfect religion, perfect policy, perfect and complete use of all things. They are even savage as we call those fruits wild which nature of herself and of her ordinary progress hath produced, whereas indeed they are those which ourselves have altered by our artificial devices and diverted from their common order we should rather term savage. In those are the true and most profitable virtues and natural proprieties most lively and vigorous which in these we have bastardized, applying them to the pleasure of our corrupted taste. And if, notwithstanding, in divers fruits of those countries that were never tilled we shall find that in respect of ours they are most excellent and as delicate unto our taste, there is no reason art should gain the point of honour of our great and puissant mother Nature. We have so much by our inventions surcharged the beauties and riches of her works that we have altogether over-choked her; yet wherever her purity shineth, she makes our vain and frivolous enterprises wonderfully ashamed.

> *Et veniunt hederae sponte sua melius,*
> *Surgit et in solis formosior arbutus antris,*
> *Et volucres nulla dulcius arte canunt.*[1]
> Ivies spring better of their own accord,
> Unhaunted plots much fairer trees afford,
> Birds by no art much sweeter notes record.

All our endeavours or wit cannot so much as reach to represent the nest of the least birdlet, its contexture, beauty, profit, and use, no, nor the web of a silly spider. 'All things', saith Plato, 'are produced either by nature, by fortune, or by art. The greatest and fairest by one or other of the two first, the least and imperfect by the last.' Those nations seem therefore so barbarous unto me because they have received very little fashion from human wit, and are yet near their original naturality. The laws of nature do yet command them, which are but little bastardized by ours. And that with such purity as I am sometimes grieved the knowledge of it came no sooner to light at what time there were men that better than we could have judged of it. I am sorry Lycurgus and Plato had it not, for meseemeth that what in those nations we see by experience doth not only exceed all the pictures wherewith licentious poesy hath proudly embellished the golden age and all her quaint inventions to feign a happy condition of man, but also the conception and desire of philosophy. They could not imagine a genuity[2] so pure and simple as we see it by experience, nor ever believe our society might be maintained with so little art and human combination. It is a nation, would I answer Plato, that hath no kind of traffic, no knowledge of letters, no intelligence of numbers, no name of magistrate nor of politic superiority, no use of service, of riches or of poverty, no contracts, no successions, no dividences,[3] no occupation but idle, no respect of

---

[1] Propertius, I.ii 10–12.
[2] Simplicity; *OED* cites only this passage.
[3] The word means both partitions and divisions of goods.

kindred but common, no apparel but natural, no manuring of lands, no use of wine, corn or metal. The very words that import lying, falsehood, treason, dissimulation, covetousness, envy, detraction, and pardon were never heard of amongst them. How dissonant would he find his imaginary commonwealth from this perfection?

> *Hos natura modos primum dedit.*[1]
> Nature at first uprise
> These manners did devise.

Furthermore, they live in a country of so exceeding pleasant and temperate situation that, as my testimonies have further assured me, they never saw any man there either shaking with the palsy, toothless, with eyes dropping, or crooked and stooping through age. They are seated alongst the seacoast, encompassed toward the land with huge and steepy mountains, having between both a hundred leagues or thereabouts of open and champaign ground. They have great abundance of fish and flesh that have no resemblance at all with ours, and eat them without any sauces or skill of cookery, but plain boiled or broiled. The first man that brought a horse hither, although he had in many other voyages conversed with them, bred so great a horror in the land that before they could take notice of him, they slew him with arrows. Their buildings are very long, and able to contain two or three hundred souls, covered with barks of great trees, fastened in the ground at one end, interlaced and joined close together by the tops after the manner of some of our granges, the covering whereof hangs down to the ground and steadeth them as a flank.[2] They have a kind of wood so hard that, riving and cleaving the same, they make blades, swords, and gridirons to broil their meat with. Their beds are of a kind of cotton cloth fastened to the house roof, as our ship-cabins; everyone hath his several couch, for the women lie from their husbands. They rise with the sun and feed for all day as soon as they are up, and make no more meals after that. They drink not at meat, as Suidas reporteth of some other people of the east which drank after meals, but drink many times a day, and are much given to pledge carouses. Their drink is made of a certain root, and of the colour of our claret wines, which lasteth but two or three days; they drink it warm. It hath somewhat a sharp taste, wholesome for the stomach, nothing heady, but laxative for such as are not used unto it, yet very pleasing to such as are accustomed unto it. Instead of bread they use a certain white composition like unto corianders confected. I have eaten some, the taste whereof is somewhat sweet and wallowish.[3] They spend the whole day in dancing. Their young men go

---

[1] Virgil, *Georgics*, ii. 20.
[2] Serves them as a wall.
[3] Insipid.

a-hunting after wild beasts with bows and arrows. Their women busy themselves therewhilst with warming of their drink, which is their chiefest office. Some of their old men, in the morning before they go to eating, preach in common to all the household, walking from one end of the house to the other, repeating one self-same sentence many times till he have ended his turn (for their buildings are a hundred paces in length), he commends but two things unto his auditory: first, valour against their enemies, then lovingness unto their wives. They never miss, for their restraint, to put men in mind of this duty, that it is their wives which keep their drink lukewarm and well-seasoned. The form of their beds, cords, swords, blades, and wooden bracelets wherewith they cover their hand-wrists when they fight, and great canes open at one end by the sound of which they keep time and cadence in their dancing, are in many places to be seen, and namely in mine own house. They are shaven all over, much more close and cleaner than we are, with no other razors than of wood or stone. They believe their souls to be eternal, and those that have deserved well of their gods to be placed in that part of heaven where the sun riseth, and the cursed toward the west in opposition. They have certain prophets and priests which commonly abide in the mountains and very seldom show themselves unto the people; but when they come down, there is a great feast prepared and a solemn assembly of many townships together (each grange as I have described maketh a village, and they are about a French league one from another). The prophet speaks to the people in public, exhorting them to embrace virtue and follow their duty. All their moral discipline containeth but these two articles: first, an undismayed resolution to war; then an inviolable affection to their wives. He doth also prognosticate of things to come and what success they shall hope for in their enterprises; he either persuadeth or dissuadeth them from war; but if he chance to miss of his divination, and that it succeed otherwise than he foretold them, if he be taken, he is hewn in a thousand pieces and condemned for a false prophet. And therefore he that hath once mis-reckoned himself is never seen again. Divination is the gift of God, the abusing whereof should be a punishable imposture. When the divines amongst the Scythians had foretold an untruth, they were couched along upon hurdles full of heath or brushwood, and so manacled hand and foot, burned to death. Those which manage matters subject to the conduct of man's sufficiency are excusable, although they show the utmost of their skill. But those that gull and cony-catch us with the assurance of an extraordinary faculty, and which is beyond our knowledge, ought to be double punished, first because they perform not the effect of their promise, then for the rashness of their imposture and unadvisedness of their fraud. They war against the nations that lie beyond their mountains, to which they go naked, having no other weapons than bows or wooden swords sharp at one end, as our broaches[1] are. It is an admirable thing to see the constant resolution of their combats, which never end but by effusion of blood and murder, for they know not what fear or routs are. Every victor brings home the head of the enemy he hath slain as a trophy of his victory and fasteneth the same at the entrance of his dwelling place. After they have long time used and entreated their prisoners well and with all commodities they can devise, he that is the master of them, summoning a great assembly of his acquaintance, tieth a cord to one of the prisoners' arms, by the end whereof he holds him fast, with some distance from him for fear he might offend him, and giveth the other arm bound in like manner to the dearest friend he hath, and both in the presence of all the assembly kill him with swords; which done, they roast and then eat him in common, and send some slices of him to such of their friends as are absent. It is not, as some imagine, to nourish themselves with it (as anciently the Scythians wont to do), but to represent an extreme and inexpiable revenge. Which we prove thus: some of them perceiving the Portugals, who had confederated themselves with their adversaries, to use another kind of death when they took them as prisoners, which was to bury them up to the middle and against the upper part of the body to shoot arrows, and then, being almost dead, to hang them up, they supposed that these people of the other world (as they who had sowed the knowledge of many vices amongst their neighbours and were much more cunning in all kinds of evils and mischief than they) undertook not this manner of revenge without cause, and that consequently it was more smartful and cruel than theirs, and thereupon began to leave their old fashion to follow this. I am not sorry we note the barbarous horror of such an action, but grieved that prying so narrowly into their faults we are so blinded in ours. I think there is more barbarism in eating men alive than to feed on them being dead; to mangle by tortures and torments a body full of lively sense, to roast him in pieces, to make dogs and swine to gnaw and tear him in mammocks[2] (as we have not only read, but seen very lately, yea, and in our own memory, not amongst ancient enemies, but our neighbours and fellow citizens, and —which is worse—under pretence of piety and religion), than to roast and tear him after he is dead. Chrysippus and Zeno, arch-pillars of the Stoic sect, have supposed that it was no hurt at all, in time of need, and to what end soever, to make use of our carrion bodies and to feed upon them as did our forefathers, who, being besieged by Caesar in the city of Alexia, resolved to sustain the famine of the siege with the bodies of old men, women, and other persons unserviceable and unfit to fight.

[1] A broach is a spear or spit.
[2] Shreds.

Vascones (fama est) alimentis talibus usi
Produxere animas.[1]
Gascons (as fame reports)
Lived with meats of such sorts.

And physicians fear not, in all kinds of compositions availful to our health, to make use of it, be it for outward or inward applications; but there was never any opinion found so unnatural and immodest that would excuse treason, treachery, disloyalty, tyranny, cruelty, and such like, which are our ordinary faults. We may then well call them barbarous in regard of reason's rules, but not in respect of us that exceed them in all kind of barbarism. Their wars are noble and generous, and have as much excuse and beauty as this human infirmity may admit: they aim at nought so much, and have no other foundation amongst them, but the mere jealousy of virtue. They contend not for the gaining of new lands, for to this day they yet enjoy that natural uberty[2] and fruitfulness which without labouring toil doth in such plenteous abundance furnish them with all necessary things that they need not enlarge their limits. They are yet in that happy estate as they desire no more than what their natural necessities direct them; whatsoever is beyond it is to them superfluous. Those that are much about one age do generally entercall[3] one another brethren, and such as are younger they call children, and the aged are esteemed as fathers to all the rest. These leave this full possession of goods in common, and without individuity to their heirs, without other claim or title but that which nature doth plainly impart unto all creatures, even as she brings them into the world. If their neighbours chance to come over the mountains to assail or invade them, and that they get the victory over them the victors' conquest is glory and the advantage to be and remain superior in valour and virtue; else have they nothing to do with the goods and spoils of the vanquished, and so return into their country, where they neither want any necessary thing nor lack this great portion, to know how to enjoy their condition happily, and are contented with what nature affordeth them. So do these when their turn cometh. They require no other ransom of their prisoners but an acknowledgement and confession that they are vanquished. And in a whole age a man shall not find one that doth not rather embrace death than either by word or countenance remissly to yield one jot of an invincible courage. There is none seen that would not rather be slain and devoured than sue for life or show any fear. They use their prisoners with all liberty that they may so much the more hold their lives dear and precious, and commonly entertain them with threats of future death, with the torments they shall endure, with the preparations intended for that purpose, with mangling and slicing of their

[1] Juvenal, Satire xv, 93.
[2] Abundance.
[3] Mutually call.

members and with the feast that shall be kept at their charge. All which is done to wrest some remiss[1] and exact some faint-yielding speech of submission from them, or to possess them with a desire to escape or run away, that so they may have the advantage to have daunted and made them afraid and to have forced their constancy. For certainly true victory consisteth in that only point.

> *Victoria nulla est*
> *Quam quae confessos animo quoque subiugat hostes.*[2]
> No conquest such as to suppress
> Foes' hearts, the conquest to confess.

The Hungarians, a most warlike nation, were whilom wont to pursue their prey no longer than they had forced their enemy to yield unto their mercy. For having wrested this confession from him, they set him at liberty without offence or ransom, except it were to make him swear never after to bear arms against them. We get many advantages of our enemies that are but borrowed and not ours: it is the quality of a porterly rascal, and not of virtue, to have stronger arms and sturdier legs; disposition is a dead and corporal quality. It is a trick of fortune to make our enemy stoop and to blear his eyes with the sun's light; it is a prank of skill and knowledge to be cunning in the art of fencing, and which may happen unto a base and worthless man. The reputation and worth of a man consisteth in his heart and will; therein consists true honour. Constancy is valour, not of arms and legs, but of mind and courage; it consisteth not in the spirit and courage of our horse, nor of our arms, but in ours. He that obstinately faileth in his courage, *si succiderit, de genu pugnat*,[3] if he slip or fall, he fights upon his knee. He that in danger of imminent death is no whit daunted in his assuredness, he that in yielding up his ghost beholdeth his enemy with a scornful and fierce look, he is vanquished not by us, but by fortune; he is slain but not conquered. The most valiant are often the most unfortunate. So are there triumphant losses in envy of victories. Not those four sister victories, the fairest that ever the sun beheld with his all-seeing eye, of Salamine,[4] of Plataea, of Mycale, and of Sicilia, durst ever dare to oppose all their glory together to the glory of the King Leonidas his discomfiture and of his men at the passage of Thermopyles: what man did ever run with so glorious an envy or more ambitious desire to the goal of a combat that Captain Ischolas to an evident loss and overthrow? Who so ingeniously or more politicly did ever assure himself of his welfare than he of his ruin? He was appointed to defend a certain passage of Peloponnesus against the Arcadians, which finding himself altogether unable to perform, seeing the

---

[1] Weakness.
[2] Claudian, *De Sexto Consulatu Honorii*, v. 248–9.
[3] Seneca, *De Providentia*, 2.
[4] Salamis.

nature of the place and inequality of the forces, and resolving that whatsoever should present itself unto his enemy must necessarily be utterly defeated; on the other side, deeming it unworthy both his virtue and magnanimity and the Lacedaemonian name to fail or faint in his charge, between these two extremities he resolved upon a mean and indifferent course,[1] which was this. The youngest and best disposed of his troop he reserved for the service and defence of their country, to which he sent them back, and with those whose loss was least and who might best be spared he determined to maintain that passage, and by their death to force the enemy to purchase the entrance of it as dear as possibly he could—as indeed it followed. For being suddenly environed round by the Arcadians, after a great slaughter made of them both himself and all his were put to the sword. Is any trophy assigned for conquerors that is not more duly due unto these conquered? A true conquest respecteth rather an undaunted resolution and honourable end than a fair escape, and the honour of virtue doth more consist in combating than in beating. But to return to our history, these prisoners, howsoever they are dealt withal, are so far from yielding that contrariwise, during two or three months that they are kept they ever carry a cheerful countenance, and urge their keepers to hasten their trial; they outrageously defy and injure them. They upbraid them with their cowardliness and with the numbers of battles they have lost against theirs. I have a song made by a prisoner wherein is this clause:

'let them boldly come together and flock in multitudes to feed on him, for with him they shall feed upon their fathers and grandfathers that heretofore have served his body for food and nourishment. These muscles [saith he] this flesh and these veins are your own; fond men as you are, know you not that the substance of your forefathers' limbs is yet tied unto ours? Taste them well, for in them shall you find the relish of your own flesh.'

An invention that hath no show of barbarism. Those that paint them dying, and that represent this action when they are put to execution, delineate the prisoners spitting in their executioners' faces and making mows[2] at them. Verily, so long as breath is in their body they never cease to brave and defy them both in speech and countenance. Surely in respect of us these are very savage men, for either they must be so in good sooth, or we must be so indeed: there is a wondrous distance between their form and ours. Their men have many wives, and by how much more they are reputed valiant, so much the greater is their number. The manner and beauty in their marriages is wondrous strange and remarkable, for the same jealousy our wives have to keep us from the love and affection of other women, the same have theirs to procure it. Being more careful for their husbands' honour and content than of anything else, they endeavour and apply all their industry to have as many rivals as possibly they

[1] A middle way.
[2] Grimaces.

can, forasmuch as it is a testimony of their husband's virtue. Our women would count it a wonder, but it is not so: it is a virtue properly matrimonial, but of the highest kind. And in the Bible, Leah, Rachel, Sarah, and Jacob's wives brought their fairest maiden servants unto their husbands' beds. And Livia seconded the lustful appetites of Augustus, to her great prejudice. And Stratonice, the wife of King Deiotarus, did not only bring a most beauteous chambermaid that served her to her husband's bed, but very carefully brought up the children he begot on her, and by all possible means aided and furthered them to succeed in their father's royalty. And lest a man should think that all this is done by a simple and servile or awe-full duty unto their custom, and by the impression of their ancient customs' authority, without discourse or judgement, and because they are so blockish and dull-spirited that they can take no other resolution, it is not amiss we allege some evidence of their sufficiency. Besides what I have said of one of their warlike songs, I have another amorous canzonet, which beginneth in this sense: 'Adder, stay, stay good adder, that my sister may by the pattern of thy particoloured coat draw the fashion and work of a rich lace for me to give unto my love; so may thy beauty, thy nimbleness or disposition be ever preferred before all other serpents'.' This first couplet is the burden of the song. I am so conversant with poesy that I may judge this invention hath no barbarism at all in it, but is altogether Anacreontic. Their language is a kind of pleasant speech, and hath a pleasing sound and some affinity with the Greek terminations. Three of that nation, ignoring how dear the knowledge of our corruptions will one day cost their repose, security, and happiness, and how their ruin shall proceed from this commerce, which I imagine is already well advanced (miserable as they are to have suffered themselves to be so cozened by a desire of newfangled novelties, and to have quit the calmness of their climate to come and see ours), were at Rouen in the time of our late King Charles the Ninth, who talked with them a great while. They were showed our fashions, our pomp, and the form of a fair city; afterward some demanded their advice, and would needs know of them what things of note and admirable they had observed amongst us: they answered three things, the last of which I have forgotten and am very sorry for it, the other two I yet remember. They said first they found it very strange that so many tall men with long beards, strong and well armed, as were about the king's person (it is very likely they meant the Switzers of his guard) would submit themselves to obey a beardless child, and that we did not rather choose one amongst them to command the rest. Secondly—they have a manner of phrase whereby they call men but a moiety of men from others[1]—they had perceived there were men amongst us full gorged with all sorts of commodities, and others which

---

[1] i.e. they speak of men as halves of each other.

were hunger-starven and bare with need and poverty begged at their gates, and found it strange these moieties so needy could endure such an injustice, and that they took not the others by the throat or set fire on their houses. I talked a good while with one of them, but I had so bad an interpreter, and who did so ill apprehend my meaning, and through his foolishness was so troubled to conceive my imaginations, that I could draw no great matter from him. Touching that point wherein I demanded of him what good he received by the superiority he had amongst his countrymen (for he was a captain, and our mariners called him king) he told me it was to march foremost in any charge of war; further, I asked him how many men did follow him; he showed me a distance of place, to signify they were as many as might be contained in so much ground, which I guessed to be about four or five thousand men; moreover I demanded if when wars were ended all his authority expired; he answered that he had only this left him, which was that when he went on progress and visited the villages depending of him, the inhabitants prepared paths and highways athwart the hedges of their woods for him to pass through at ease. All that is not very ill, but what of that? They wear no kind of breeches or hosen.

From 'Of Cruelty', *Book 2, Chapter 11 (p. 243)*

He that through a natural facility and genuine mildness should neglect or contemn injuries received should no doubt perform a rare action, and worthy commendation. But he who, being touched and stung to the quick with any wrong or offence received, should arm himself with reason against this furiously-blind desire of revenge, and in the end, after a great conflict, yield himself master over it, should doubtless do much more. The first should do well, the other virtuously : the one action might be termed goodness, the other virtue.

MEDEA'S INCANTATION

FROM Ovid, *Metamorphoses*, vii. 179–219

Tres aberant noctes, ut cornua tota coirent
efficerentque orbem; postquam plenissima fulsit
ac solida terras spectavit imagine luna,
egreditur tectis vestes induta recinctas,
nuda pedem, nudos umeris infusa capillos,
fertque vagos mediae per muta silentia noctis
incomitata gradus: homines volucresque ferasque
solverat alta quies, nullo cum murmure saepes,
inmotaeque silent frondes, silet umidus aer,
sidera sola micant: ad quae sua bracchia tendens
ter se convertit, ter sumptis flumine crinem
inroravit aquis ternisque ululatibus ora
solvit et in dura submisso poplite terra
'Nox' ait 'arcanis fidissima, quacque diurnis
aurea cum luna succeditis ignibus astra,
tuque, triceps Hecate, quae, coeptis conscia nostris
adiutrixque venis cantusque artisque magorum,
quaeque magos, Tellus, pollentibus instruis herbis,
auraeque et venti montesque amnesque lacusque,
dique omnes nemorum, dique omnes noctis adeste,
quorum ope, cum volui, ripis mirantibus amnes
in fontes rediere suos, concussaque sisto,
stantia concutio cantu freta, nubila pello
nubilaque induco, ventos abigoque vocoque,
vipereas rumpo verbis et carmine fauces,
vivaque saxa sua convulsaque robora terra
et silvas moveo iubeoque tremescere montis
et mugire solum manesque exire sepulcris
te quoque, Luna, traho, quamvis Temesaea labores
aera tuos minuant; currus quoque carmine nostro
pallet avi, pallet nostris Aurora venenis
vos mihi taurorum flammas hebetastis et unco
inpatiens oneris collum pressistis aratro,
vos serpentigenis in se fera bella dedistis
custodemque rudem somni sopistis et aurum
vindice decepto Graias misistis in urbes:

nunc opus est sucis, per quos renovata senectus
in florem redeat primosque recolligat annos,
et dabitis. neque enim micuerunt sidera frustra,
nec frustra volucrum tractus cervice draconum
currus adest.' aderat demissus ab aethere currus.

From *The xv. Bookes of P. Ouidius Naso, entytuled Metamorphosis, translated oute of Latin into English meeter, by Arthur Golding Gentleman.* . . (1567)

Before the moon should circlewise close both her horns in one
Three nights were yet as then to come. As soon as that she shone
Most full of light, and did behold the earth with fulsome face,
Medea with her hair not trussed so much as in a lace,
But flaring on her shoulders twain, and barefoot, with her gown
Ungirded, got her out of doors and wandered up and down
Alone the dead time of the night. Both man and beast and bird
Were fast asleep; the serpents sly in trailing forward stirred
So softly as you would have thought they still asleep had been.
The moisting air was whist; no leaf ye could have moving seen.
The stars alonely fair and bright did in the welkin shine.
To which she lifting up her hands did thrice herself incline,
And thrice with water of the brook her hair besprinkled she,
And gasping thrice she oped her mouth, and bowing down her knee
Upon the bare, hard ground she said, 'O trusty time of night
Most faithful unto privities[1], O golden stars whose light
Doth jointly with the moon succeed the beams that blaze by day,
And thou three-headed Hecatè, who knowest best the way
To compass this our great attempt and art our chiefest stay;
Ye charms and witchcrafts, and thou earth, which both with herb and weed
Of mighty working furnishest the wizards at their need;
Ye airs and winds; ye elves of hills, of brooks, of woods alone,
Of standing lakes, and of the night, approach ye every one,
Through help of whom (the crooked banks much wond'ring at the thing)
I have compellèd streams to run clean backward to their spring.
By charms I make the calm seas rough and make the rough seas plain,
And cover all the sky with clouds and chase them thence again.
By charms I raise and lay the winds and burst the viper's jaw,
And from the bowels of the earth both stones and trees do draw.
Whole woods and forests I remove; I make the mountains shake,
And even the earth itself to groan and fearfully to quake.
I call up dead men from their graves; and thee, O lightsome moon,

---

[1] Secret matters.

I darken oft, though beaten brass abate thy peril soon;
Our sorcery dims the morning fair and darks the sun at noon.
The flaming breath of fiery bulls ye quenchèd for my sake,
And causèd their unwieldy necks the bended yoke to take.
Among the earth-bred brothers you a mortal war did set,
And brought asleep the dragon fell whose eyes were never shet,[1]
By means whereof deceiving him that had the golden fleece
In charge to keep, you sent it thence by Jason into Greece.
Now have I need of herbs that can by virtue of their juice
To flowering prime of lusty youth old withered age reduce.
I am assured ye will it grant; for not in vain have shone
These twinkling stars, ne yet in vain this chariot all alone
By draught of dragons hither comes.' With that was from the sky
A chariot softly glancèd down, and stayèd hard thereby.

[1] Shut, an old past participle.

APPENDIX F

LINEATION

In cases where a verse line is divided between two speakers, the Folio regularly prints the second part of the line at the left-hand margin. Since Steevens's time it has been normal editorial practice to indicate the relation of the two parts by indenting the second. Such instances are not recorded here as departures from the Folio lineation.

1.1. 56–7	This . . . tides] POPE; *as prose* F
60–2	CAPELL; *as verse, dividing after* 'us' *and* 'children' F
1.2.304	POPE; *two lines, dividing after* 'hence' F
309–10	'Tis . . . on] POPE; *one line in* F
360	THEOBALD; F *ends the line with* 'hadst', l. 361
378–9	DOVER WILSON; *one line in* F
380–7	*Hark . . . dow*] CAPELL; *four lines dividing after* 'bark', 'bow-wow', *and* 'Chanticleer' F
2.1. 13–14	POPE; *as verse, dividing after* 'wit' F
29–30	POPE; *as verse, dividing after* 'wager' F
53	POPE; *two lines, dividing after* 'looks' F
190	POPE; F *divides after* 'thoughts'
193–4	It . . . comforter] ROWE 1714; *one line in* F
194–6	We . . . safety] ROWE 1714; F *divides after* 'person'
199–200	Doth . . . sleep] ROWE 1714; F *divides after* 'find'
242–3	Then . . . Naples] POPE; *one line in* F
304–5	Now . . . King] STAUNTON; *one line in* F
2.2. 41–2	CAPELL; *one line in* F
43–4	POPE; *as verse, dividing after* 'man's' F
54	POPE; *two lines, dividing after* 'too' F
56	What's . . . here] POPE; *two lines, dividing after* 'matter' F
83	I . . . be] POPE; *two lines, dividing after* 'voice' F
114	POPE; *two lines, dividing after* 'scape' F
124	POPE; *a separate line in* F
129–30	How . . . ague] POPE; *a separate line in* F
139–41	The . . . sooth] POPE; *as verse, dividing after* 'moon' *and* 'monster' F
152–3	POPE; *as verse, dividing after* 'drink' F
154–5	POPE; *as prose* F
157–8	POPE; *as prose* F
161–6	POPE; *as prose* F

	176–7	CAPELL; *one line in* F
3.2.	38–9	POPE; *as verse, dividing after* 'it' F
	40–2	POPE; *as verse, dividing after* 'tyrant' *and* 'me' F
	47–8	POPE; *as verse, dividing after* 'tale' F
	57–8	POPE; *two lines, dividing after* 'compassed' F
	67	Trinculo . . . danger] POPE; *a separate line in* F
	70	POPE; *two lines, dividing after* 'nothing' F
	73	Do . . . that] POPE; *a separate line in* F
	76	A . . . do] POPE; *a separate line in* F
	106–7	Dost . . . Trinculo] POPE; *a separate line in* F
	109–10	POPE; *as verse, dividing after* 'thee' F
	117–18	POPE; *as verse, dividing after* 'do reason' F
	119–21	GLOBE; *two lines, dividing after* 'and flout 'em' F
	129–30	POPE; *as verse, dividing after* 'thee' F
	142–3	This . . . nothing] POPE; *as verse, dividing after* 'me' F
	145	POPE; *two lines, dividing after* 'and by' F
	146–7	POPE; *as verse, dividing after* 'away' F
	148–9	POPE; *as verse, dividing after* 'monster' *and* 'taborer' F
	150	POPE; *two lines, dividing after* 'come' F
3.3.	13 4	The . . . thoroughly] POPE; *one line in* F
4.1.	196–200	POPE; *as verse, dividing after* 'fairy', 'which' *and* 'should' F
	210–13	POPE; *as verse, dividing after* 'wetting', 'monster' *and* 'bottle' F
	219–22	POPE; *as verse, dividing after* 'hand', 'thoughts' *and* 'Stephano' F
5.1.	95	I . . . thee] F3; F1 *divides the line after* 'miss'
	220	POPE; *two lines, dividing after* 'land' F
	256–8	POPE; *as verse, dividing after* 'let' *and* 'is' F
	278	POPE; *two lines, dividing after* 'now' F
	282–4	POPE; *as verse, dividing after* 'last' *and* 'bones' F

INDEX

This is a guide to words defined in the Commentary and to a selection of names and topics in the Introduction. An asterisk indicates that the note supplements information given in *OED*.

Index

mudded, 3.3.102; 5.1.151
muse, *v.* 3.3.36

naiads, 4.1.128
nature/nurture, pp. 24, 28;
    4.1.188-9
Neoplatonic, pp. 11, 20, 23
Nobody, 3.2.124-5
nymphs, 4.1.132

oared, 2.1.116
operatic versions, pp. 66-7, 68, 69,
    70, 74-5
Ovid, *Metamorphoses*, pp. 19, 20, 41,
    App. E

pageant, 4.1.155
*patch, 3.2.62
perdition, 1.2.30; 3.3.77
performance at court, pp. 1, 2, 58,
    61-2
phoenix, 3.3.23
pickle, 5.1.281
piece, 1.2.56
pig-nuts, 2.2.162
pioned, 4.1.64
pipe (musical instrument), 3.2.122.1
pitch, 1.2.3
plantation, 2.1.141
plot, 3.2.106
poll-clipped, 4.1.68
productions: 17th-c., pp. 1, 2, 3, 61,
    64-6, 71-2; 18th-c., pp. 66-9, 72,
    75; 19th-c., pp. 69-71, 72, 73, 75,
    76; 20th-c., pp. 73-5, 76-87
project, 2.1.297; 5.1.1
Proserpine, 5.1.88-9, 114-15
Prospero: pp. 14-20; 1.2.0.1;
    18th-c., pp. 6-7; 19th-c., pp. 7,
    10; 20th-c., pp.11-12, 13; age, pp.
    50, 55, 79-83; and Caliban, pp.
    23-8, 83; and James I, pp. 1, 4,
    21; and magic, pp. 14, 16, 20-3,
    37, 50, 51-4; and masque, pp. 20,
    21, 43-50, 61-2; name, pp. 42-3;
    and Shakespeare, pp. 1, 4 10, 12,
    63, 80; and Sycorax, pp. 20-3;
    theatrical interpretations, pp.
    79-86; and wife, pp. 16, 18,
    19-20, 38-9
proverbial expressions: 1.1.28-30,
    52; 1.2.30, 94, 214, 335, 441-2,
    470; 2.1.25, 29-32, 34, 136, 151,

152-3, 158, 180-2, 219-20;
    2.2.56-7, 57, 59-60, 68, 78-9,
    93-4, 124; 3.1.89-90; 3.2.18, 33,
    49-50, 60, 69, 110,
    121,127-7,129; 3.3.26, 39, 82.2;
    4.1.51-2, 164, 197, 206, 239,
    262; 5.1.238, 281
Purcell, Henry, p. 66
putter-out, 3.3.48
psychoanalytic criticism, pp. 11-12,
    13, 19, 37-8

quality, 1.2.193

race, 1.2.357
rack, 4.1.156
raven, 1.2.321-2
reconciliation, pp. 5, 10, 13, 18, 29,
    30, 47, 51, 53, 54
renunciation, pp. 13, 50-3, 55
revels, 4.1.148
Revels Accounts, pp. 1, 2, 3, 62
roarers, 1.1.17
*rounded, 4.1.158

savages, 2.2.57
*scamels, 2.2.166
Sebastian, 1.1.8.1
sense, 2.1.104-5
Setebos, p. 33; 1.2.372; 5.1.261
Shadwell, Thomas, pp. 7, 66, 67, 68,
    69, 72, 74
signories, 1.2.71
single, 1.2.433
sparrows, 4.1.100
spirits/sprites, 2.2.112
standard, 3.2.16
stomach, 1.2.156-7; 2.1.104-5
Strachey, William, letter, pp. 32,
    62-3, 89, App.B
storm scene, pp. 14-15
subtleties, 5.1.124
sweet, 4.1.124
Sycorax, pp. 19-20, 21, 25; 1.2.258

tabor, 3.2.122.1; 4.1.175
temple, 1.2.458
text: date, pp. 62-4; editorial
    procedures, p. 89; Folio, position in,
    pp. 1, 4, 56, 58-9; as playscript, p.
    12; revision, theories of, pp. 1 4,
    61-2; stage directions, pp.57-8;
    traditional cuts, pp. 16-17, 18, 72,
    76; transcribed, pp. 12, 26, 56-8;
    (*see also* Ralph Crane)

247